Intermediate 2
CHEMISTRY

Norman Conquest

Hodder & Stoughton

A MEMBER OF THE HODDER HEADLINE GROUP

Acknowledgements

Action Plus (83 bottom, 122 both, 123 top, 127 bottom, 128 top left, 141 top); AKG (2 bottom, 195); Andrew Lambert (129, 155 bottom); Corbis (3 top, 30, 83 middle, 131, 171, 192, 203); Hodder & Stoughton (204); Holt Studios (174); Life File (7 both, 12 top, 51, 84 top right, 130, 141 bottom, 169, 170, 172, 206 bottom right); Norman Conquest (3 bottom, 9 both, 11, 12 bottom, 15 both, 25, 26 all, 27, 38, 42 all, 45, 81, 82, 83 top, 84 top left and bottom, 89 both, 100, 102, 103, 110, 111 both, 113, 115, 123 bottom, 124, 125 all, 126, 127 top, 128 top right, 137, 140 bottom three, 155 top, 160 both, 182, 200, 206 bottom left); PA Photos (5); Roddy Paine (76); Ruth Hughes (140 top, 145, 149); Science & Society Picture Library (183); Science Photo Library (135) /Martin Land (196) /NASA (2 top, 6); Still Pictures (173)

Orders: please contact Bookpoint Ltd, 130 Milton Park, Abingdon, Oxon OX14 4SB. Telephone: (44) 01235 827720, Fax: (44) 01235 400454. Lines are open from 9.00–6.00, Monday to Saturday, with a 24 hour message answering service. Email address: orders@bookpoint.co.uk

British Library Cataloguing in Publication Data
A catalogue record for this title is available from The British Library

Published by Hodder & Stoughton Educational Scotland

ISBN 0 340 78205 6

First published 2001
Impression number   10 9 8 7 6 5 4 3 2 1
Year   2006 2005 2004 2003 2002 2001

ISBN 0 340 78206 4

First published 2001
Impression number   10 9 8 7 6 5 4 3 2 1
Year   2006 2005 2004 2003 2002 2001

With Answers

Printed in Great Britain for Hodder & Stoughton Educational, a division of Hodder Headline Plc, 338 Euston Road, London NW1 3BH by The Bath Press Ltd.

Section 1
Building Blocks

Section 2
Carbon Compounds

Section 3
Acids, Bases and Metals

Section 4
Whole Course Summaries and Questions

The text of this book follows closely the course structure of the Higher Still Intermediate 2 Chemistry course, with in-text questions given to help candidates understand what has just been covered. The three main units have been sub-divided into chapters corresponding to sub-divisions (a), (b), etc. in the course content statements. Most chapters end with Study Questions provided in the style of the NAB end-of-unit tests.

A series of whole course summaries has been provided, with related questions provided for the majority of them. A chemical dictionary has also been included. As end-of-unit questions tend to be quite narrow in their scope, a considerable number of whole course questions, taken from SQA and SEB past papers, have been provided to reflect the more challenging nature of the national examination.

Prescribed Practical Activities have been described at appropriate points in the text with sufficient details provided to satisfy the requirements of the external examination. A selection of questions has been provided in the final section in order to test knowledge and understanding of the nine PPAs.

Acknowledgements

I wish to acknowledge the encouragement given to me by colleagues and former colleagues at Perth Academy. I am also indebted to my editors at Hodder & Stoughton for their patience and meticulous attention to detail, particularly Ruth Hughes and Stephen Halder. Grateful and much overdue acknowledgement is also due to my wife, Anne, for her help and support during the production of this and previous texts.

Norman Conquest 2001

Building Blocks

1 Substances

Elements

Figure 1.1 ▲
Everything in or on the Earth is made from about one hundred elements.

Figure 1.2 ▲
Gold is an element which occurs uncombined in the Earth's crust. It was used to make Tutankhamun's funeral mask.

Everything that you see around you is made from about one hundred **elements** that are the basic building blocks of the whole world. They are the simplest kind of substance and cannot be broken down into anything simpler. Hydrogen and oxygen are examples of elements.

Just as many different things can be made from a few basic pieces of Lego, so, millions of different substances can be made by the joining together of relatively few elements. Substances that are formed by the chemical joining of elements are called **compounds**. The elements hydrogen and oxygen, for example, can join together to form the compound that we know as water.

Some elements, like gold, have been known for thousands of years, whereas others have been discovered more recently. Aluminium, for example, is the most plentiful metal in the Earth's crust, but it was only discovered as recently as 1827. By the early part of the twentieth century, most of the naturally occurring elements had been discovered. Then, from 1939 onwards, a series of elements were made by scientists that did not occur naturally.

The elements are arranged in a special way in the **Periodic Table**. Most of the first ninety-two elements occur naturally, although usually joined with other elements. Each element is given a **symbol**, which can be a single letter, for example O for oxygen and H for hydrogen. In other cases two letters are used, for example Ca for calcium and Mg for magnesium. In some cases the symbol is derived from a foreign language, for example Fe is the symbol for iron because the Latin word for iron is *ferrum*.

Most of the elements are solids at room temperature but a few are gases and two, mercury and bromine, are liquids. The gases include hydrogen, oxygen, nitrogen, fluorine, chlorine and a family of very unreactive elements called the **noble gases**.

Chemists find it very useful to classify elements as being either **metals** or **non-metals**. A typical metal is a shiny electrical conductor, whereas a typical non-metal is a dull non-conductor. Most of the elements are metals.

The Periodic Table places elements with similar chemical properties in the same vertical column. A column of elements in the Periodic Table is called a **group**. Some of the groups are given names.

Figure 1.3 ▲
Dimitri Mendeléev, a brilliant Russian chemistry teacher who, in 1869, invented the Periodic Table based on the 60 elements then known. (He cut his hair once a year, in spring, when the weather started to get warmer.)

Three important groups of the Periodic Table are:

Group 1 – **alkali metals** (lithium, sodium, potassium, rubidium, caesium, francium)
Group 7 – **halogens** (fluorine, chlorine, bromine, iodine, astatine)
Group 0 – **noble gases** (helium, neon, argon, krypton, xenon, radon)

The alkali metals are similar in that they are all very reactive. The halogens are reactive non-metals, while the noble gases are all extremely unreactive.

In between groups 2 and 3 there is a large block of elements called **transition metals**. They also have similar properties, one of which is that they tend to form coloured compounds when they join with other elements. The transition metals include some that are well known, such as iron, copper, zinc and silver.

Horizontal rows of elements in the Periodic Table show a change in properties as one goes across the table (see page 4). On the left, in group 1, are the very reactive alkali metals. In the centre are less reactive metals and non-metals, but in group 7 are the very reactive halogens, followed by the very unreactive non-metal noble gases on the right in group 0. The horizontal rows of elements are called **periods**.

Questions

1 Use the data booklet to help you answer the following:
   (a) Which three noble gases were discovered in 1898?
   (b) Which group 3 metal would melt in your hand?
   (c) Name two alkali metals which float on water.
   (d) Which two elements are liquids at room temperature (25°C)?

2 (a) Give chemical symbols for copper, zinc, tin and lead. (Refer to the Periodic Table on page 4.)
   (b) Which two transition metals were known to the Romans as *aurum* and *argentum*?
   (c) Other than appearance, what test would you carry out to discover whether an element was a metal or a non-metal?
   (d) Find out what the names 'hydrogen' and 'halogen' mean.

Compounds

Relatively few elements are found uncombined. Oxygen, nitrogen and the noble gases make up most of the air and some solid elements such as gold, sulphur and carbon are found in the Earth's crust. Most of the substances in our world are either pure compounds or mixtures of compounds, mainly the latter. For example, sand is made up mainly of silicon dioxide, a compound of silicon and oxygen. Water is another common compound, made up of hydrogen and oxygen and having the chemical name, hydrogen oxide.

Figure 1.4 ▲
Sand is mainly silicon dioxide and the sea is mainly water (hydrogen oxide). Both silicon dioxide and hydrogen oxide are very common compounds.

The Periodic Table of Elements

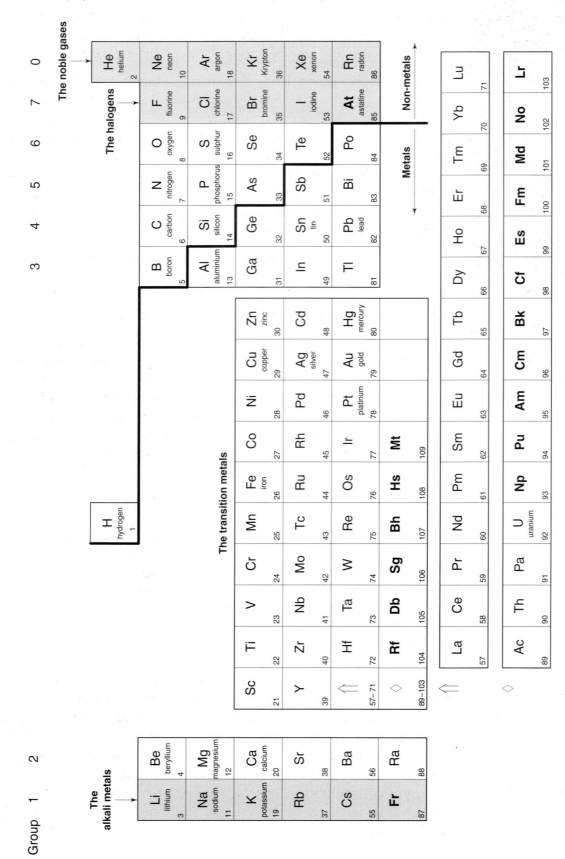

Group 1 2   3 4 5 6 7 0

Figure 1.5 ▲
During firework displays metals join with oxygen to produce metal oxide compounds giving out light of various colours.

Many metal elements form compounds readily with non-metal elements, especially when heated together. The result can be quite spectacular, with a lot of heat and light being released. Metal powders, like magnesium and iron for example, are used in fireworks.

Compounds have completely different properties from the elements that make them up. The magnesium oxide that is produced when magnesium combines with oxygen is a white powder, whereas magnesium is a silvery solid and oxygen is a colourless gas. The white crystalline compound, sodium chloride, which is also known as 'common salt' is also very different from the elements in it. Sodium is a soft silvery reactive metal and chlorine is a poisonous green gas. When *two* elements react to form a compound, the name of the compound ends in **–ide**. For example,

Magnesium and oxygen react to give magnesium ox**ide**.

Sodium and chlorine react to give sodium chlor**ide**.

Copper and sulphur react to give copper sulph**ide**.

When a metal reacts with a non-metal, it is the ending of the non-metal which is changed to **–ide** when naming the compound formed. When two non-metals react, it is the ending of the more reactive one which is changed when naming the compound. Using this rule explains why, when hydrogen and oxygen react, the compound is called hydrogen oxide and not oxygen hydride. Some more examples of compounds containing two elements are given in Table 1.1.

Name of compound	Elements present
Potassium chloride	potassium and chlorine
Aluminium bromide	aluminium and bromine
Silver iodide	silver and iodine
Copper oxide	copper and oxygen
Mercury sulphide	mercury and sulphur

Table 1.1 ▲
Some compounds containing two elements.

Many compounds contain more than two elements, with oxygen being present in many of them. Chemists have a special way of showing that oxygen is present in such a compound. The ending of the name for the compound is either **–ate** or **–ite** and usually three elements are present, one of which is oxygen. In magnesium sulphate, for example, the elements present are magnesium, sulphur and oxygen. In magnesium sulphite, the same elements are present, but there is less oxygen than in magnesium sulphate. Some more examples are given in Table 1.2.

Questions

3 Give the names of the elements present in the following compounds: **(a)** tin iodide; **(b)** nickel sulphide; **(c)** lead oxide; **(d)** magnesium nitride; **(e)** calcium carbide.

4 Give the names of the elements present in the following compounds: **(a)** barium sulphate; **(b)** sodium nitrite; **(c)** potassium chlorate; **(d)** zinc carbonate; **(e)** magnesium phosphate.

Name of compound	Elements present
Calcium carbonate	calcium, carbon and oxygen
Potassium nitrate	potassium, nitrogen and oxygen
Lithium phosphate	lithium, phosphorus and oxygen
Sodium sulphate	sodium, sulphur and oxygen
Sodium sulphite	sodium, sulphur and oxygen

Table 1.2 ▲
Some compounds containing three elements.

Figure 1.6 ▲
*Clouds of the compound steam form
when hydrogen and oxygen combine
readily in a space shuttle's main motors —
but it is not easy to reverse the process.*

**The test for oxygen is that it
relights a glowing splint.**

Air does not give this result
because it does not contain
enough oxygen.

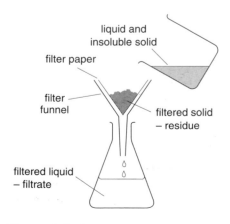

liquid and
insoluble solid

filter paper

filter
funnel

filtered solid
– residue

filtered liquid
– filtrate

Figure 1.7 ▲
Apparatus for filtration.

It is usually not easy to separate a compound into its elements again.
For example, when hydrogen and oxygen combine to produce water,
as happens in the main rocket motor of a space shuttle, the water
produced does not readily turn back into hydrogen and oxygen.
Heating water strongly simply causes it to boil producing steam, which
is still water but in the form of a gas; it does not break up water to
give hydrogen and oxygen again.

Another feature of compounds is that their composition is fixed. In
water, for example, there is always 88.9% oxygen by mass and 11.1%
hydrogen.

Mixtures

A **mixture** is formed when two or more substances, which can be
elements or compounds, are mixed together but do not join
chemically. Most substances that occur naturally are mixtures. Air, for
example, is a mixture of elements and compounds. It consists mainly
of elements, with nitrogen and oxygen being the most plentiful, but
compounds such as carbon dioxide and water vapour are also present.
Crude oil is a very complicated mixture, made up of many
compounds.

Most mixtures must be separated before they can be used and
fortunately, unlike compounds, they are usually easy to separate. In
order to separate a solid from a liquid, such as mud from a sample of
muddy water, **filtration** is used. The solid left behind in the filter
paper is called the **residue**, which in this case is the mud. The liquid
that passes through the filter paper is called the **filtrate**, which is the
water in this case.

If we want to separate a dissolved solid from a liquid, then
evaporation can be used. Using this technique, sea salts can be
separated from sea water and pure water can be collected from the
steam that is driven off. The harvesting of salt from the sea is now
confined to hotter countries, but used to take place in cooler countries
as well, including Scotland.

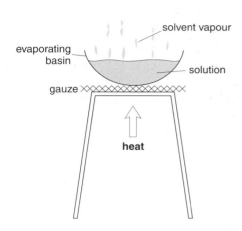

Figure 1.8 ▲
Apparatus for evaporation.

Figure 1.9 ▲
In hot countries salt is obtained from sea water by using the sun's rays to evaporate off the water.

A mixture of liquids can be separated by **distillation** if they have different boiling points. The mixture is heated and the liquid with the lowest boiling point will boil first and can be condensed and collected. During whisky making, ethanol, which boils at 79°C, can be separated from water, which boils at 100°C. During distillation, the liquid collected is called the **distillate**. The glass apparatus used in the school laboratory is rather different from the copper pot stills used during whisky making which are shown on in Figure 1.10.

Even tiny quantities of dissolved substances can be separated by **chromatography** using a piece of absorbent paper. Food dyes are sold as solutions that are often mixtures of different coloured compounds.

Figure 1.10 ▲
Copper stills used to make whiskey.

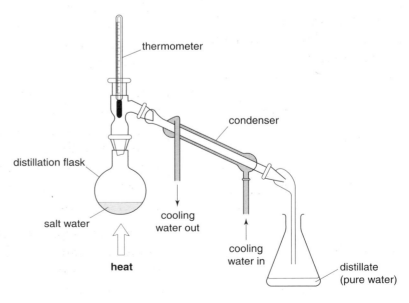

Figure 1.11 ▲
Apparatus for distillation.

7

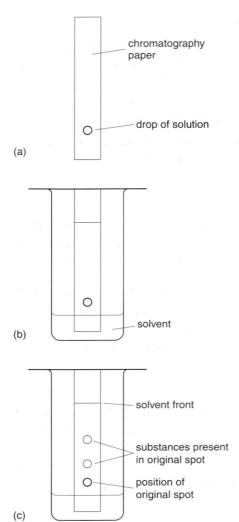

chromatography paper

drop of solution

(a)

solvent

(b)

solvent front

substances present in original spot

position of original spot

(c)

Figure 1.12 ▲
Paper chromatography can work using only a single drop of a solution.

The mixture can be separated by putting a small spot of dye on the chromatography paper (Figure 1.12a) and dipping the end of the paper in a suitable liquid, such as water or the alcohol ethanol (Figure 1.12b). The liquid slowly moves up through the spot of dye, separating it into its different components as it goes (Figure 1.12c). A green food dye, for example, may be found to consist of two compounds, one blue and the other yellow. Why not try the experiment at home using a piece of absorbent paper such as blotting paper?

The word chromatography was used to describe this technique because the first substances to be separated were coloured and the Greek word for colour is *chroma*. Nowadays, the substances being separated may be colourless, such as sugars and amino acids. In these cases the paper is sprayed with a liquid which reacts with the colourless substances to produce coloured substances. In this way their positions on the paper can be seen.

Questions

5 Some processes used in chemistry to separate mixtures are: distillation; filtration; evaporation; chromatography.
   Which process would you use:

   (a) to find out how many coloured substances are present in an orange dye;
   (b) to obtain sugar from a sugar solution;
   (c) to obtain insoluble chalk powder from a mixture of chalk and water;
   (d) to obtain pure water from copper sulphate solution.

6 State how you would separate a mixture of common salt and sand to obtain dry samples of each.

Compound	Mixture
A single substance in which two or more elements are chemically joined.	Two or more substances (elements or compounds) mixed together but not chemically joined.
The properties are different from those of the elements in it.	The properties are similar to those of the substances in it.
The composition is always the same.	The composition can vary.
The elements in it can only be separated by a chemical reaction.	The substances in it are easily separated by a physical process.

Table 1.3 ▲
Differences between compounds and mixtures.

Figure 1.13 ▲
A sugar solution about to be made – but do you like yours dilute or concentrated?

Solvents, solutes and solutions

When sugar dissolves in tea or coffee it disappears; we can't see it, but it must still be there because we can taste it. A special kind of mixture called a **solution** has been formed. Solutions are very common because many substances will dissolve in water. In all cases, the substance that dissolves is called the **solute** and the liquid in which the solute dissolves is called the **solvent**. A **dilute solution** contains only a little solute compared to the amount of solvent, whereas a **concentrated solution** contains a lot of solute compared to the amount of solvent.

Solute particles in a solution are so small that they cannot be removed by filtration as they pass straight through the pores of filter paper. The only way to retrieve a solid solute is to evaporate off the solvent.

Water is a very good solvent for many solutes, but there are still many substances that do not dissolve in it. Nail varnish, for example, is insoluble in water but dissolves readily in propanone, which is sometimes referred to by its older name of acetone. Nail varnish remover often contains propanone.

When a substance like sugar dissolves in a liquid such as water, then we say that the substance is **soluble**. Substances which do not dissolve are said to be **insoluble**. Sand, for example is insoluble in water. With all solutes there tends to be a limit as to how much can be dissolved in a given amount of solvent. When no more solute will dissolve in a solvent, then the solution is said to be **saturated**. Usually, solutes are more soluble in hot solvents than cold ones.

The change in solubility with temperature can be shown by a solubility curve. From this it can be seen that 100 g of water can dissolve 20 g of copper sulphate at 20°C and that the solubility of copper sulphate increases appreciably with rising temperatures.

Figure 1.14 ▲
Nail polish remover is a solvent for nail polish. This one contains propanone.

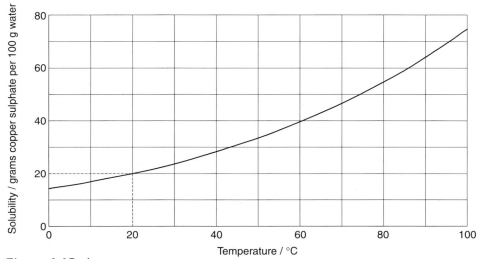

Figure 1.15 ▲
Solubility curve for copper sulphate.

Questions

Use Figure 1.15 to answer these questions.

7 What is the solubility of copper sulphate at 60°C?

8 A saturated solution of copper sulphate in 100 g of water is cooled from 50°C to 30°C.

   (a) What is the solubility of copper sulphate at 30°C?
   (b) What mass of copper sulphate crystals is thrown out of solution?

The graph shows the mass of copper sulphate in grams that is needed to produce a saturated solution, using 100 g of water, at each temperature.

The solubility curve in Figure 1.15 can be used to predict what will happen when a hot saturated solution of copper sulphate cools down. It is possible to dissolve 55 g of copper sulphate in 100 g of water at 80°C. But when the solution cools to 20°C, only 20 g of solute can remain dissolved. The other 35 g of copper sulphate are thrown out of the solution and appear as crystals. This process of crystallisation can be used to produce crystals of different sizes. If a hot saturated solution cools quickly, then small crystals are produced, but if cooling is slower then larger crystals are formed.

Sugar manufacturers make use of this information when producing sugar crystals. Caster sugar is made up of small crystals, which are produced on fast cooling, whereas granulated sugar crystals are larger and are the result of slower cooling.

State symbols

State symbols can be used to indicate the state of a substance:

(s) = solid
(l) = liquid
(g) = gas
(aq) = aqueous (dissolved in water)

We can use state symbols when dealing with the process of dissolving. For example, when solid sugar dissolves in water we can write:

$$\text{sugar(s)} + \text{aq} \rightarrow \text{sugar(aq)}$$

This tell us that sugar and water produce a sugar solution.

$$+ = \text{'and'}$$

$$\text{aq} = \text{'water'}$$

$$\rightarrow = \text{'produce'}$$

When liquid ethanol dissolves in water we can write:

$$\text{ethanol(l)} + \text{aq} \rightarrow \text{ethanol(aq)}$$

Carbon dioxide gas is present in fizzy drinks, so when it dissolves in water to form a solution, the dissolving process is shown as:

$$\text{carbon dioxide(g)} + \text{aq} \rightarrow \text{carbon dioxide(aq)}$$

Chemical reactions and physical changes

How can we tell whether a chemical reaction has taken place? The answer lies in deciding whether a new substance has been formed or not. This is because during a **chemical reaction** one or more new substances are always formed. There may be other clues as well because in many chemical reactions there is a detectable energy change and there is often a change in appearance as well. The burning of wood or coal are good examples of chemical reactions as they involve all three things that we should look for. New substances are formed, including smoke, gases and ash. There is a definite energy change, as we can feel the heat given off and see the light. The appearance changes too, from the log or piece of coal to ash, smoke and gases.

Physical changes take place when *no* new substance is formed. There may, however, be energy changes and a change in appearance. For example, a current of electricity can cause a metal wire to glow, indicating both an energy change and a change in appearance. But since no new substance has been formed, this is not a chemical reaction. All changes of state are also classified as physical changes for the same reason. When ice melts and produces liquid water, both the ice and the liquid water are made up of the same substance, namely water, so although there has been a change in appearance a physical change has taken place, not a chemical reaction.

Another important difference between chemical reactions and physical changes is that it is usually quite difficult to reverse a chemical reaction, whereas physical changes are usually easily reversed. Liquid water is easily frozen to produce solid ice again, but it is hard to imagine how smoke, gases and ash could turn back into wood and coal.

Question

9 Explain why a candle burning is an example of a chemical reaction, but wax melting is not.

Figure 1.16 ▲
Hill walkers and mountaineers use the exothermic reaction from toe warmers as a source of heat energy.

Chemical reaction	Physical change
New substance(s) formed	No new substance(s) formed
Detectable energy change in many	Detectable energy change in some
Change in appearance in many	Change in appearance in many
Usually difficult to reverse	Usually easy to reverse

Table 1.4 ▲
Chemical reactions and physical changes compared.

Exothermic and endothermic energy changes

When sodium hydroxide dissolves in water, there is a very noticeable rise in temperature; the solution becomes very hot if enough sodium hydroxide is used. Energy changes like this, where heat energy is given out and the temperature rises, are said to be **exothermic**. Some toe warmers, used by hill walkers and mountaineers rely on exothermic reactions to provide a source of heat. A more widely used chemical reaction which releases heat energy is **burning**, which chemists also call **combustion**.

Figure 1.17 ▲
The exothermic reactions that take place when fuels burn are widely used as a source of heat energy.

The dissolving of ammonium nitrate in water results in the solution cooling down as heat energy is taken in by the chemicals. 'Cold packs', which are used to treat some sports injuries, contain ammonium nitrate and water in two compartments, separated by a thin membrane. When the pack is squeezed, the membrane breaks, and the ammonium nitrate dissolves in the water, accompanied by a drop in temperature. Energy changes, where heat energy is taken in and the temperature falls, are described as **endothermic**. Endothermic reactions are not as common as exothermic ones, but you can try out a perfectly safe one yourself. Simply put some sherbet in your mouth. The mixture fizzes and the temperature drops a little because the reaction is endothermic. The reaction taking place is between citric acid and sodium hydrogencarbonate (bicarbonate of soda). The dry powders do not react, but on coming into contact with water they do.

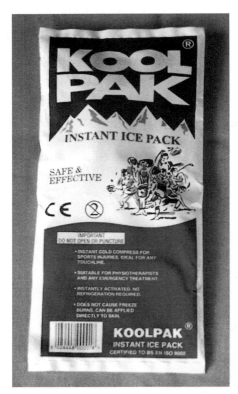

Figure 1.18 ▲
If you pull a muscle, one of these might be put on the affected area.

In questions 1–4 choose the correct word **from the following list** to complete the sentence.

endothermic, 3, solvent, period, solute, 2, exothermic, group

1 A column of elements in the Periodic Table is called a _____.

2 A substance that dissolves in a liquid to form a solution is called a _____.

3 A reaction that takes in energy from its surroundings causing a drop in temperature is said to be _____.

4 Compounds with the ending **–ite** or **–ate** usually contain _____ elements, one of which is oxygen.

5 The correct symbol for copper is
A Co
B Ce
C Cu
D Cr

6 Which of the following is an alkali metal?
A Calcium
B Sulphur
C Aluminium
D Lithium

7 Which of the following is a halogen?
A Phosphorus
B Bromine
C Krypton
D Nitrogen

8 Which of the following is a noble gas?
A Nitrogen
B Oxygen
C Fluorine
D Neon

9 Which of the following is a transition metal?
A Nickel
B Lead
C Strontium
D Magnesium

10 Between which groups of the Periodic Table are the transition metals found?
A 1 and 2
B 2 and 3
C 3 and 4
D 4 and 5

11 Which of the following elements is a metal?
A Arsenic
B Selenium
C Tellurium
D Bismuth

12 Which of the following has a melting point of $232°C$?
A Neon
B Sulphur
C Astatine
D Tin

13 Which of the following elements have similar chemical properties?
A Sulphur and helium
B Calcium and strontium
C Sodium and germanium
D Iron and nitrogen

14 Which of the following elements was discovered first?
A Iodine
B Zinc
C Nickel
D Oxygen

15 At very low temperatures, air becomes liquid. Liquid nitrogen is made from liquid air by a process of distillation.

Why can liquid air be separated in this way?

16 Which noble gas was discovered in 1894?

17 A sample of air was found to have the following composition by volume:
Nitrogen   78%
Oxygen   21%
Other gases   1%

Present this information as a bar graph.

18 A line graph showing how the solubility of potassium chlorate changes with temperature is shown below:

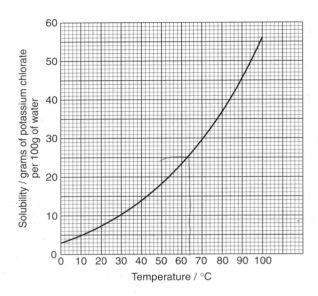

Temperature / °C

(a) What is the maximum mass of potassium chlorate that would dissolve in 100 g of water at 62°C?

(b) At what temperature is the solubility of potassium chlorate 15 g per 100 g of water?

(c) Calculate the mass of crystals that would form if a solution containing 80 g of potassium chlorate in 200 g of water were cooled from 90°C to 30°C.

19 Some orange ammonium dichromate crystals were placed on a heat-resistant mat and touched with a hot spatula. Immediately, a green powder was thrown up into the air, as if from a volcano. The centre of the substance glowed red hot for about 20 seconds, after which all that was left was the green powder. Give *three* pieces of evidence which suggest that a chemical reaction has taken place.

20 Complete the following tables:

(a)

Compound	Elements present	
hydrogen fluoride	hydrogen	fluorine
tin oxide	_____	_____
iron _____	_____	sulphur
calcium nitride	_____	_____
lithium hydride	_____	_____
_____ _____	sodium	iodine

(b)

Compound	Elements present		
calcium nitrate	calcium	nitrogen	oxygen
lead sulphate	_____	_____	_____
zinc sulphite	_____	_____	_____
lithium carbonate	_____	_____	_____
_____ phosphate	barium	_____	_____
_____ nitrite	potassium	_____	_____

21 Normal air does not relight a glowing splint, but when dissolved air is recovered from water, the mixture of gases obtained does relight a glowing splint.

(a) Which gaseous element relights a glowing splint?

(b) What conclusion can you reach regarding the solubilities of the two main gases present in air?

22 A company makes 'Sparkling spring water' by dissolving carbon dioxide in water under pressure.

Use the words below to complete the sentence.

solution, solvent, solute

In order to make 'Sparkling spring water', water, the _____, has carbon dioxide gas, the _____, forced into it under pressure to produce 'Sparkling spring water', a _____ of carbon dioxide in water.

23 (a) Long ago, sea water was used in Scotland as a source of salt. How was the salt extracted from the sea water?

(b) The chemical industry in North East England grew, initially, due to the discovery of salt beneath the Cheshire plain. One method of extraction is to mine it directly, but another method is now favoured which leaves behind insoluble impurities.
Describe a possible alternative method for obtaining salt from these underground deposits.

② Reaction Rates

Factors affecting rate of reaction

When a piece of magnesium ribbon is lit using a bunsen flame it can take several seconds to burn completely. If the same mass of magnesium powder is poured into the flame, the burning is much quicker and the reaction is over in less than a second. This example illustrates an important factor that affects the rate of reactions involving solid reactants. A reaction can be speeded up by *decreasing the particle size* of any solid reactant.

What would happen if you took two identical bottles of milk, but left one at room temperature for several days and put the other in the refrigerator for the same length of time? The chemical reactions that occur when milk goes sour take place faster at a higher temperature. This is generally true of all reactions. A reaction can be speeded up by *increasing the temperature* at which the reaction occurs.

When a reaction involves a solution it is found that changing the concentration of the solution alters the rate of reaction. For example, if identical marble chips are placed in hydrochloric acid of different concentrations, carbon dioxide gas is given off faster with the more concentrated acid. In general, a reaction can be speeded up by *increasing the concentration* of a reactant in solution.

The things that we can change during a chemical reaction, such as particle size and concentration, are called **variables.** When investigating the effect on reaction of changing a variable, such as concentration of the acid in Figure 2.3, all other variables *must* be kept constant. The comparison is then fair.

Figure 2.1 ▲
Magnesium ribbon burns quite quickly, but not as fast as magnesium powder!

Figure 2.2 ▲
Why must we store milk at a low temperature?

Question

1   In Figure 2.3, what variables must be kept constant in order to make the comparison fair?

Summary

Chemical reactions are speeded up by

- **decreasing the particle size** of any solid reactant
- **increasing the temperature** at which the reaction occurs
- **increasing the concentration** of a reactant in solution

When investigating the effect of change in a **variable**, only *one* should be changed at a time.

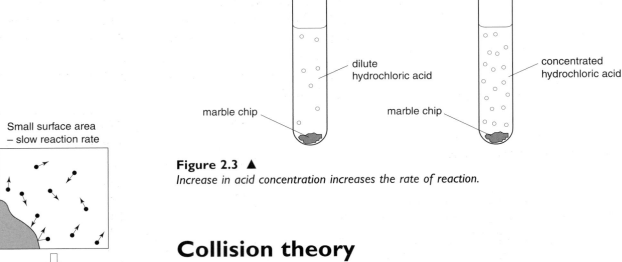

Figure 2.3 ▲
Increase in acid concentration increases the rate of reaction.

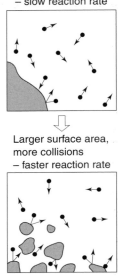

Small surface area
– slow reaction rate

Larger surface area,
more collisions
– faster reaction rate

Figure 2.4 ▲

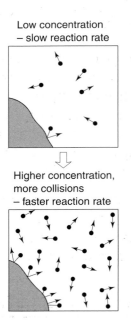

Low concentration
– slow reaction rate

Higher concentration,
more collisions
– faster reaction rate

Figure 2.5 ▲

Collision theory

All matter is made up of very small particles, which in later topics will be described as atoms, ions or molecules. For there to be a possibility of a chemical reaction occurring, these particles must collide. Two solids are therefore unlikely to react, since the particles present can only vibrate about fixed positions. Reactions may occur if at least one of the substances present has particles that are free to move about and therefore be involved in collisions. Gases and liquids, including solutions, have particles that are free to move about and might therefore take part in chemical reactions.

Small marble chips react faster with dilute hydrochloric acid than an equal mass of larger chips because they have a greater surface area. There can be more collisions, in a given period of time, between the acid particles and the smaller chips and it is this which causes the greater rate of reaction. Figure 2.4 illustrates these points.

Concentrated hydrochloric acid reacts faster with marble chips than dilute hydrochloric acid for a similar reason. Although the surface area for collision is now the same, there is an increase in the concentration of acid particles in solution and, as a result, more collisions between acid particles and the surface of the marble in any given period of time. This is illustrated in Figure 2.5.

Not all collisions between particles result in a reaction taking place. When the gas is turned on before lighting a bunsen burner, for example, there is no reaction between the gas and oxygen in the air despite the fact that their particles are definitely colliding. Particles need a certain minimum energy before they can react and, in the case of natural gas and oxygen in the air, this is usually provided by a spark or by a flame.

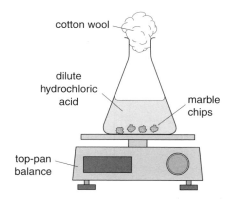

cotton wool

dilute hydrochloric acid

marble chips

top-pan balance

Figure 2.6 ▲
Measuring the rate of reaction between dilute hydrochloric acid and marble chips.

Following the course of a reaction

Reactions between liquids and/or solids in which gases are given off are easy to follow. One such example is the reaction between marble chips and an acid such as dilute hydrochloric acid. As the reaction takes place, the flask containing the marble and acid becomes lighter as carbon dioxide escapes. The reaction can be followed by putting the flask containing the reaction mixture on a balance and noting the loss in mass at regular time intervals (see Figure 2.6). Some typical results are shown in Table 2.1.

Time/s	Mass of carbon dioxide produced/g
0	0.00
30	1.25
60	1.92
90	2.40
120	2.76
150	3.06
180	3.32
210	3.52
240	3.68
270	3.81
300	3.92
360	4.11
420	4.23
480	4.30
540	4.35
600	4.35

Table 2.1 ▲
The results can be plotted as a line graph, with time on the horizontal axis and the mass of carbon dioxide produced on the vertical axis (see Figure 2.7).

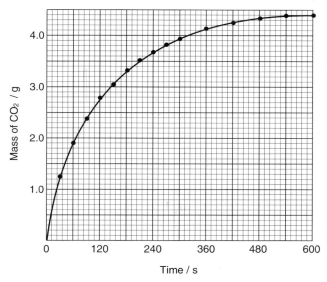

Figure 2.7 ▲
Graph of mass of carbon dioxide produced against time.

The rate of reaction at any moment in time is given by the slope of the graph at that time. The slope of the graph in Figure 2.7 decreases as the reaction proceeds. As there is little change in temperature during this reaction, the decrease in rate of reaction is partly due to a decrease in acid concentration and, probably to a lesser extent, a decrease in the surface area of the marble chips. Both of these factors reduce the collision rate between the acid and the marble.

It is easier to measure the *average* rate of reaction over a period of time than to measure the rate at a particular moment in time since the latter is always changing. For our example, where the mass of carbon dioxide produced is being monitored, the average rate of reaction is given by:

$$\text{Average rate of reaction} = \frac{\text{change in mass of carbon dioxide}}{\text{change in time}}$$

Using the results obtained, 1.92 g of carbon dioxide was produced in the first 60 seconds. We can calculate the average rate of reaction over the first 60-second interval as follows:

$$\text{Average rate of reaction during first 60 seconds} = \frac{1.92}{60} = 0.032\,\text{g s}^{-1}$$

During the second 60-second interval, the mass of carbon dioxide given off was $(2.76 - 1.92) = 0.84\,\text{g}$.

$$\text{Average rate of reaction during second 60 seconds} = \frac{0.84}{60} = 0.014\,\text{g s}^{-1}$$

It can be seen from these two calculations that the average rate of reaction decreases as the reaction proceeds.

Many other quantities may be monitored in order that a reaction can be followed and the average rate of reaction calculated. For example, in the reaction between an acid and marble chips, it is possible to measure the *volume* of carbon dioxide gas produced at time intervals. This can be done either by using a gas syringe, or by collecting the gas over water in an upturned measuring cylinder (see Figure 2.8). It is also possible to monitor the decrease in concentration of a reactant, such as hydrochloric acid, or the increase in concentration of a product. Whatever quantity is measured during the course of a reaction, the average rate of reaction over a certain period of time is always given by:

$$\textbf{Average rate of reaction} = \frac{\textbf{change in quantity monitored}}{\textbf{change in time}}$$

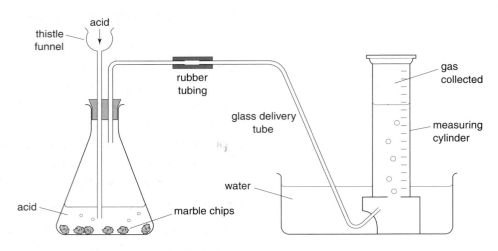

Figure 2.8 ▲
Monitoring the volume of gas produced is another way of following a reaction.

Question

2 The graph below shows how the volume of hydrogen gas given off changed with time during a reaction between pieces of zinc and dilute sulphuric acid.

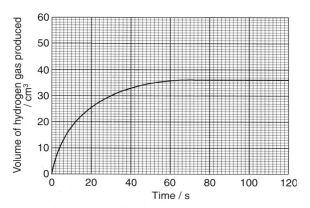

(a) Calculate the average rate of reaction over the first 20-second period.

(b) Calculate the average rate of reaction over the second 20-second period.

(c) Explain the difference in rate of reaction between (a) and (b).

Prescribed Practical Activity

Effect of concentration on reaction rate

Iodine is produced in the reaction between solutions of potassium iodide and sodium persulphate. If starch is present in the solution the iodine forms a dark blue colour with this. However, if sodium thiosulphate is also present, no blue colour is seen until all of the sodium thiosulphate has reacted with the iodine being produced. It is not possible to measure the mass of iodine produced before the appearance of the blue colour, but an experiment can be devised in which the time is noted for a constant mass of iodine to be formed.

As Figure 2.9 shows, sodium persulphate solution and starch solution are mixed in a beaker placed on a white surface. A solution containing potassium iodide and sodium thiosulphate is added and timing started. The time taken for the mixture to suddenly go dark blue is noted. The experiment is repeated using smaller volumes of the sodium persulphate solution, but adding water so that the total volume of the reacting mixture is always the same.

Continued ➤

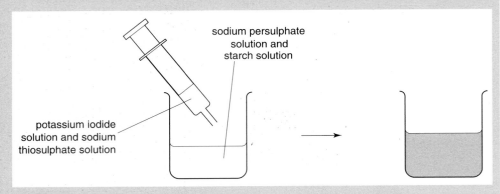

Figure 2.9 ▲
After a while the mixture suddenly becomes dark blue.

The concentrations and volumes of all other solutions is kept constant. Since the same mass of sodium thiosulphate is used in each experiment, the same mass of iodine must be produced *before* the sudden appearance of the blue colour. For this stage of the reaction we can write:

$$\text{(Average) rate of reaction} = \frac{\text{mass of iodine produced}}{\text{time taken}} = \frac{\text{a constant}}{\text{time taken}} = \frac{k}{t}$$

$$\text{(Average) rate of reaction} \propto \frac{1}{t}$$

The rate of reaction is proportional to the reciprocal of time taken (1/t). This is true for all reactions or stage in a reaction. It should also be noted that, if the rate of reaction is high, the time taken will be small, but if the rate of reaction is low, the time taken will be large.

Specimen results are given in Table 2.2 and a graph of 1/t (a measure of reaction rate) against volume of sodium persulphate solution is shown in Figure 2.10.

Experiment	1	2	3	4
Volume of sodium persulphate solution/cm³	10	8	6	4
Volume of deionised water/cm³	0	2	4	6
Time for dark blue colour to appear (t)/s	44	54	71	107
Reaction rate $(1/t)/s^{-1}$	0.0227	0.0185	0.0141	0.0093

Table 2.2 ▲

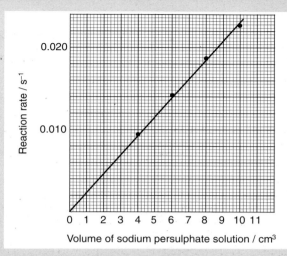

Figure 2.10 ▲

The graph of reaction rate against volume of sodium persulphate is a straight line. Since the total volume of solution is always the same, we can take the volume of sodium persulphate solution used to be a measure of its concentration. The line graph therefore tells us that the reaction rate is *directly proportional* to the concentration of the sodium persulphate.

> A **variable** is something that can be changed in a chemical reaction, e.g. temperature, particle size, volume, concentration, etc.

In any experiment of this kind it is important that only one **variable** is changed, in this case the volume of the sodium persulphate solution (which directly affects its concentration in the reaction). All other variables must be kept constant so that the comparisons of reaction rate are fair and valid. These other variables include the total volume of the mixture, the volume and concentration of the potassium iodide solution (containing sodium thiosulphate) and the temperature of the mixture.

Prescribed Practical Activity

Effect of temperature on reaction rate

In the reaction between solutions of hydrochloric acid and sodium thiosulphate, a suspension of finely divided sulphur is produced which eventually makes the mixture so opaque that it is impossible to see through it. It is not possible to measure the mass of sulphur produced in a given time, but an experiment can be devised in which the time is noted for a constant mass of sulphur to be formed.

Continued ➤

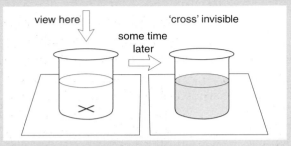

Figure 2.11 ▲

As Figure 2.11 shows, a beaker containing 20 cm³ of sodium thiosulphate solution is placed over a cross, lightly drawn in pencil, on a piece of paper. 1 cm³ of hydrochloric acid is added, the mixture is stirred and timing started. Viewing from above, the time for the disappearance of the cross is noted. The temperature of the mixture is measured before washing and drying the beaker. The experiment is repeated, but with the sodium thiosulphate solution being heated and its temperature raised by about a further 10°C each time before use. It is, however, the final mixture temperature that is noted for each experiment. Since the same mass of sulphur is produced in each experiment by the time the cross is no longer visible, for this stage of the reaction we can write:

Safety note

1 The sulphur dioxide that is formed during the reaction is toxic and can cause an asthmatic attack.

2 Do not exceed 50°C in this experiment.

$$\text{(Average) rate of reaction} = \frac{\text{mass of sulphur produced}}{\text{time taken}} = \frac{\text{a constant}}{\text{time taken}} = \frac{k}{t}$$

$$\text{(Average) rate of reaction} \propto \frac{1}{t}$$

Specimen results are given in Table 2.3 and a graph of '1/t' (a measure of reaction rate) against temperature is shown in Figure 2.12.

Temperature/°C	Time (t)/s	Reaction rate (1/t)/s⁻¹
20	80	0.0125
32	40	0.0250
39	27	0.0370
48	16	0.0625

Table 2.3 ▲

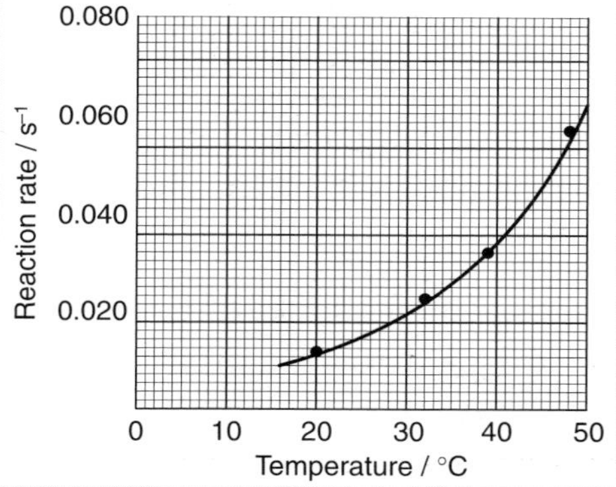

Figure 2.12 ▲

In this experiment, the variable changed is the temperature of the mixture, so all other variables must be kept constant. These include the volume and concentration of the hydrochloric acid and also the volume and concentration of the sodium thiosulphate solution. However, since the depth of mixture should be kept constant, we should use the same clean beaker each time. The same pencilled cross should be used with, ideally, the same person making the observations because eyesight varies from one person to another.

Catalysts

There is a way to speed up a chemical reaction that does not involve either increasing the temperature or concentration of a reactant, or decreasing particle size. That is, by adding a substance called a **catalyst**. Figure 2.13 shows the effect of adding manganese(IV) oxide powder to hydrogen peroxide solution. Hydrogen peroxide solution decomposes very slowly at room temperature to give water and oxygen, as shown by the equation:

$$\text{hydrogen peroxide(aq)} \rightarrow \text{water(l)} + \text{oxygen(g)}$$

Normally this reaction is so slow that you cannot see any signs of it taking place. However, when manganese(IV) oxide is added, bubbles of oxygen gas are produced very quickly and the mixture warms up. We can say that manganese(IV) oxide is a *catalyst* for the decomposition of hydrogen peroxide.

Chemists have discovered many different catalysts. They all speed up some reactions and can be recovered unchanged at the end of the reaction. They do, however, take a very active part in the reactions that they catalyse. Many catalysts are either transition metals or, like manganese(IV) oxide, are compounds containing a transition metal.

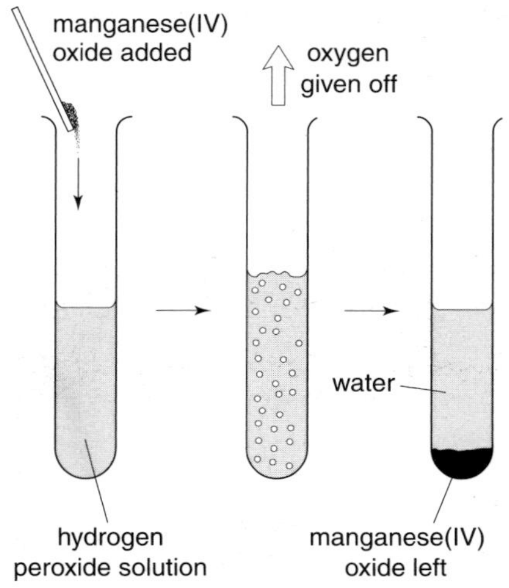

Figure 2.13 ▲
Using a catalyst to speed up a reaction.

Question

3 If exactly 1 g of manganese(IV) oxide had been used as catalyst in the reaction above, how could you show, at the end of the reaction, that 1 g of it was still there?

Two types of catalyst

Manganese(IV) oxide is said to be a **heterogeneous** catalyst for the decomposition of hydrogen peroxide solution because it is in a *different* state from the reactants. By far the most common heterogeneous catalysts are finely divided solids. They are used in this form because it provides a large surface area from a small mass of catalyst. It is on the surface of heterogeneous catalysts that reaction takes place.

Heterogeneous catalysts have *active sites* on their surface and reactant molecules form weak bonds with the surface there in a process called **adsorption** (see Figure 2.14). At the same time bonds within the adsorbed reactant molecules are weakened. The adsorbed molecules are also held at a favourable angle for a collision with another reactant molecule to be successful and result in a reaction (see Figure 2.15). The product molecules then leave the active site, in a stage called desorption, and the active site is available again for the adsorption of another reactant molecule (see Figure 2.16).

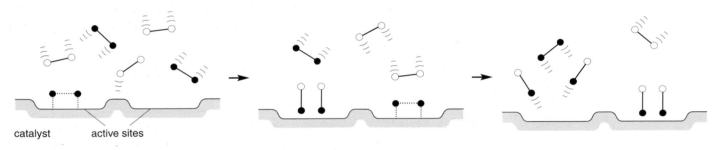

catalyst   active sites

Figure 2.14 ▲
Adsorption.

Figure 2.15 ▲
Reaction.

Figure 2.16 ▲
Desorption.

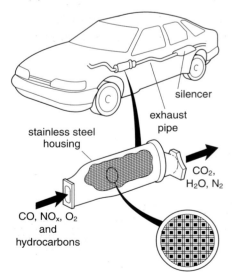

silencer

exhaust pipe

stainless steel housing

CO_2, H_2O, N_2

CO, NO_x, O_2 and hydrocarbons

ceramic honeycomb structure covered with platinum and other transition metals

Figure 2.17 ▲
A catalytic converter.

The transition metals platinum, palladium and rhodium act as heterogeneous catalysts in the catalytic converters fitted in car exhaust systems (see Figure 2.17). The effect of them, when hot, is to turn harmful and polluting gases into less harmful ones. Among the reactions that they catalyse are:

carbon monoxide + nitrogen oxides → carbon dioxide + nitrogen

unburned hydrocarbons + oxygen → carbon dioxide + water

If unwanted substances are preferentially adsorbed onto the active sites, which are then not available for the normal reactants, the catalyst is said to have been **poisoned**. When cars were first fitted with catalytic converters it was important to use only unleaded petrol. The reason for this was that some cars needed leaded petrol, which was available until 1999, but the lead present poisoned the catalysts in the catalytic converters.

Figure 2.18 ▲
The combustion inside this styling brush is catalysed by platinum.

The heterogeneous catalysts aluminium oxide and aluminium silicate are used when cracking hydrocarbons in the oil industry. The active sites get covered up by carbon during the process and the catalyst loses its efficiency. Fortunately the catalyst can be regenerated by constantly removing some of it from the reaction chamber and burning off the carbon. Sometimes it is not possible to regenerate a poisoned catalyst and it must be renewed. This adds to industry's costs, so every effort is made to remove any impurities from reactants that might poison a catalyst.

An everyday example of a heterogeneous catalyst is found in some heated styling brushes, where finely divided platinum is used to catalyse the reaction between the hydrocarbon fuel methylpropane and oxygen in the air. The fuel burns, heating up the styling brush.

A catalyst that is in the same state as the reactants is said to be a **homogeneous** catalyst. Many of these are solutions, but there are examples where catalyst and reactants are all gases. In reactions of this type, the catalyst forms an intermediate compound with one of the reactants, which decomposes to reform the catalyst at a later stage. In a simple example where reactants **A** and **B** react to form products, this is what might happen:

$$\text{catalyst} + \textbf{A} \rightarrow \text{intermediate compound}$$

then, intermediate compound + **B** → products + catalyst

This is what happens when cobalt(II) ions in aqueous solution catalyse the reaction between aqueous solutions of hydrogen peroxide and potassium sodium tartrate. When a mixture of these two solutions is heated hydrogen peroxide molecules and tartrate ions react slowly to release oxygen gas. Addition of an aqueous solution of cobalt(II) chloride, which contains pink cobalt(II) ions, causes the reaction to speed up dramatically.

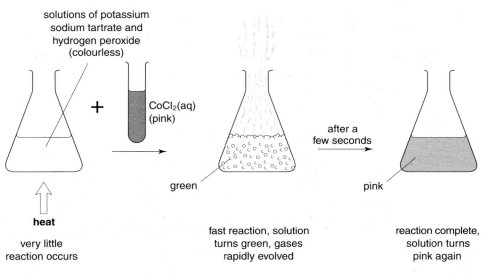

Figure 2.19 ▶
Reaction between potassium sodium tartrate and cobalt(II) chloride.

The frothing mixture turns green due to the formation of cobalt(III) ions as part of an intermediate compound, before turning pink again, as cobalt(II) ions are reformed when the reaction ends and frothing subsides. The sequence of changes that take place is shown in Figure 2.19.

Enzymes

Enzymes are biological catalysts. They catalyse the chemical reactions that take place in the living cells of plants and animals. The human body contains over 30 000 different enzymes; the reason for so many is that enzymes are *specific*, meaning that they can only catalyse one reaction, or at best one type of reaction. One of these enzymes, called catalase, helps to break down poisonous hydrogen peroxide into water and oxygen. Catalase is also present in the liver of other animals and in plant tissues. As Figure 2.20 shows, ox liver can decompose hydrogen peroxide very rapidly.

The enzyme invertase catalyses the break down of the sugar sucrose into two simpler sugars, glucose and fructose. It is used industrially in the production of soft-centred chocolates. The effect of invertase on sucrose can be investigated using Benedict's solution. A solution of sucrose does not affect the Benedict's solution when the two are heated together. However, if a little invertase is added to the sucrose solution and the mixture is kept warm for a minute or two, the resulting mixture turns the blue Benedict's solution orange. This happens because both glucose and fructose react with Benedict's solution. The reaction catalysed by invertase may be written as:

$$\text{sucrose} + \text{water} \rightarrow \text{glucose} + \text{fructose}$$

Unlike other catalysts, enzymes are easily destroyed by heat. Raising the temperature to about 70°C is sufficient to destroy most enzymes.

Figure 2.20 ▲
Liver contains the enzyme catalase, which speeds up the decomposition of hydrogen peroxide.

Figure 2.21 ▲
Enzymes are important in the production of bread, cheese and wine.

For example, cooked liver has no effect on hydrogen peroxide and if invertase is heated to about 70°C, it no longer has any effect on sucrose. Body enzymes work best close to body temperature, 37°C. Enzymes are also sensitive to changes in pH (see Section 3) and lose their activity if placed in very acidic or very alkaline conditions.

Enzymes are added to some washing powders because they help to break down biological stains such as blood and fat. Enzymes are also used in the manufacture of cheese and yoghurt. They are also used to tenderise meat and to remove haze in chilled beer. Yeast provides a variety of enzymes which are used in making bread, wine, whisky, beer and lager.

Catalysts and industry

Catalysts are of great importance industrially because they reduce the energy used in a reaction and may reduce the reaction time. The effect of both of these is to reduce costs and therefore make a manufacturing process more economical. If more than one reaction is possible from a given set of reactants, a catalyst may be available which favours one particular route. Some catalysts which are used industrially are shown in Table 2.4, along with details of the reactants, products and the manufacturing process they are relevant to.

Catalyst	Reactants	Products	Process
Aluminium oxide	Larger hydrocarbons	Smaller hydrocarbons	Cracking
Iron	Nitrogen + hydrogen	Ammonia	Making ammonia
Nickel	Vegetable oil + hydrogen	Fat	Making margarine
Invertase	Sucrose + water	Glucose + fructose	Making soft-centred chocolates
Zymase (in yeast)	Glucose	Ethanol (alcohol) + carbon dioxide	Making alcoholic drinks

Table 2.4 ▲

Chapter 2 Study Questions

In questions 1–4 choose the correct word **from the following list** to complete the sentence.

> **heterogeneous, particle size, adsorption, homogeneous, temperature, poisoning, concentration, enzyme**

1 A reaction can be slowed down by increasing the _____ of any solid reactant.

2 The reaction between the gases nitrogen and hydrogen that is catalysed by solid iron is an example of _____ catalysis.

3 A catalyst which speeds up a reaction that takes place in living cells is called an _____.

4 When reactant molecules occupy active sites on a solid catalyst the process is called _____.

5 The following compounds are all used as catalysts. Which one does **not** contain a transition metal?
A Chromium(III) oxide
B Manganese(IV) oxide
C Aluminium(III) oxide
D Vanadium(V) oxide

6 A group wanted to investigate the effect of changes in acid concentration on the reaction with zinc metal. Which of the following need **not** be kept constant?
A The mass of zinc
B The temperature of the acid
C The size of the test tube
D The particle size of the zinc

7 You are asked to measure accurately the volume of carbon dioxide produced every minute when chalk and acid react together.

Which of the following combinations of pieces of equipment (shown in the diagram) is the **best** one to use?
A P and Q
B S and Q
C P and R
D S and R

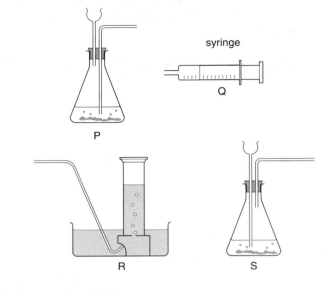

8 Graph X was obtained when 1 g of calcium carbonate powder reacted with excess dilute hydrochloric acid at 20°C.

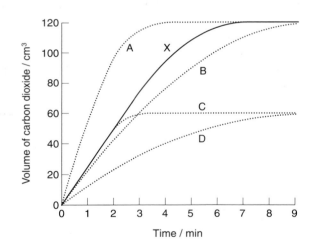

Which curve would best represent the reaction of a 0.5 g lump of calcium carbonate with excess of the same dilute hydrochloric acid?

9 The reaction between some marble chips (an excess) and dilute hydrochloric acid was followed by noting the loss in mass as carbon dioxide was given off. The following table of results was obtained.

Time/min	Mass of carbon dioxide/g
0	0
1	1.50
2	2.50
3	3.05
4	3.40
5	3.59
6	3.67
7	3.70
8	3.70

(a) Draw a line graph to show these results.
(b) Calculate the average rate of reaction during the first two minutes.
(c) Why does the reaction rate slow down?
(d) Why does the graph eventually become horizontal?

10 When hydrogen peroxide decomposes it releases oxygen gas. The reaction can be speeded up by the use of a catalyst.

Anne and Simon investigated the reaction using the apparatus below:

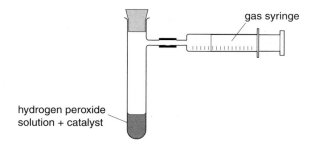

Their experiments were the same, except that Anne used manganese dioxide as the catalyst, producing graph A, whereas Simon used copper oxide, which gave graph B.

(a) State *three* factors that Anne and Simon had to keep the same in their experiments so that the comparison of the catalysts was fair.
(b) Which of the two oxides is the better catalyst?
(c) Calculate the average rate of reaction over the first 25 seconds for each reaction.

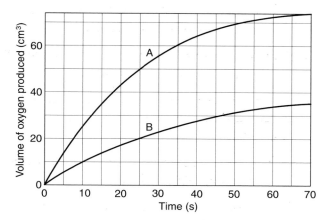

11 Many solid industrial catalysts are used as either fine powders spread out on an unreactive material, or as very fine wires in the form of a gauze if they are metals.

Why are the catalysts used in these forms, rather than as large pieces?

12 Powdered nickel is used as a catalyst for the reaction between hydrogen gas and hot vegetable oil in the production of margarine, which is liquid when first made.

(a) What type of catalysis takes place during this reaction?
(b) Suggest how the nickel could be separated from the products in this reaction?

13 In oxy-acetylene burners, the fuel gas acetylene burns in pure oxygen.

Why is this flame hotter than one where acetylene burns in air, of which only one-fifth is made up of oxygen?

The Structure of the Atom

Atoms

Every element is made up of very small particles called **atoms**. An atom is the smallest part of an element that can take part in chemical reactions. Each element is made from its own type of atom. For example, gold is made of gold atoms and helium from helium atoms. Up until the end of the nineteenth century it was believed that atoms were solid, like tiny snooker balls, then in 1897, J J Thomson, a professor of physics at Cambridge University, discovered a particle *inside* the atoms of a gas. He found that if a very high voltage was applied to a gas, very light, negatively charged particles were pulled away from the atoms of the gas. He knew that they were negatively charged because they were attracted towards a positive electrode. These negatively charged particles, from inside the atom, Thomson called **electrons**.

Since atoms were known to be electrically neutral, Thomson suggested that they must contain the same number of positively charged particles, which were later called **protons**, as negatively charged electrons. He pictured the atom as a solid sphere of positive charge (the protons), with an equal number of electrons embedded in it. This became known as the 'plum pudding' model of the atom.

However, the 'plum pudding' model of the atom did not explain the results of experiments carried out by Geiger and Marsden at Manchester University under the supervision of Professor Ernest Rutherford. They fired positively charged alpha particles at thin metal foils. Most went straight through undeflected, some were deflected a little, but a tiny number came backwards off the foil. In 1911 Rutherford published an explanation of the results of Geiger and Marsden's experiments. He proposed that atoms were mainly empty space, but that they possessed a tiny, dense, positively charged central part, which he called the **nucleus**. He said that the nucleus was where the positively charged protons were to be found and the negatively charged electrons were arranged outside the nucleus. Figure 3.2 shows how Rutherford explained Geiger and Marsden's experiments. Rutherford succeeded J J Thomson as Professor of Physics at Cambridge University.

Mathematical calculations by Thomson showed that protons were about 1850 times heavier than electrons and modern techniques show that the radius of the nucleus is about one ten-thousandth of the radius of a typical atom. Rutherford was also a brilliant mathematician and he calculated that there had to be another kind of particle present in the nucleus of almost all elements' atoms.

Figure 3.1 ▲
Ernest Rutherford who was first to propose that atoms have a nucleus.

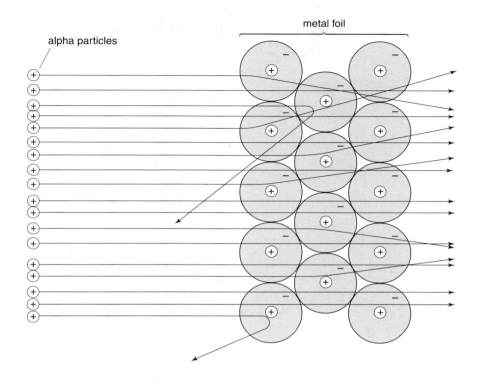

Figure 3.2 ▲
Most alpha particles pass through a metal foil, some are deflected, but a few rebounded from the foil – Ernest Rutherford explained why.

He called this particle, that had no charge but a mass equal to that of a proton, the **neutron**. Eventually, in 1932, the neutron was discovered by James Chadwick, one of Rutherford's colleagues.

Particle	Charge	Relative mass	Where found
proton	+1	1	nucleus
neutron	0	1	nucleus
electron	−1	0 (almost)	outside nucleus

Table 3.1 ▲
Protons, neutrons and electrons – a summary.

Atoms and the Periodic Table

The atoms of any particular element all have the same number of protons and it is this number, called the **atomic number**, which is also the numbered position of the element in the Periodic Table. For example, hydrogen is element number one in the Periodic Table, so its atoms have just one proton in their nuclei and the atomic number of hydrogen is 1. Iron is element number 26, so iron has an atomic number of 26 and all iron atoms have 26 protons in their nuclei.

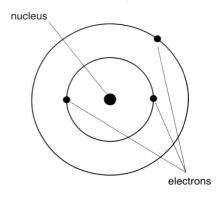

Figure 3.4 ▲
A lithium atom.

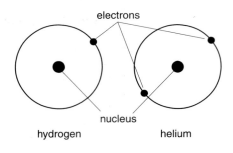

Figure 3.3 ▲
A hydrogen atom and a helium atom.

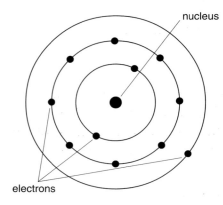

Figure 3.5 ▲
A sodium atom.

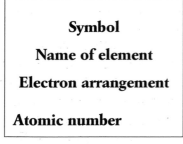

for example

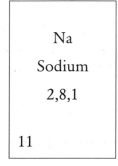

Figure 3.6 ▲

Electrons are arranged around the nucleus in special layers called **energy levels**, or **shells**. The further away from the nucleus the shells are, the more electrons they can hold. For the elements with relatively few electrons in their atoms, the shells nearest to the nucleus fill up before any electrons enter an outer shell. For example, hydrogen atoms have one electron in the first shell and helium atoms have two (see Figure 3.3). Lithium atoms have three electrons, but the first shell can only hold two electrons, so the third electron enters the second shell as shown in Figure 3.4. The electron arrangement for lithium is written as (2,1). From lithium across to neon, in the second period of the Periodic Table, electrons enter the second shell until it is full with eight electrons. The electron arrangement for neon is therefore (2,8). The next element in the Periodic Table is sodium, with 11 electrons in each atom. Since the first two shells are full, the eleventh electron enters the third shell (see Figure 3.5) and the electron arrangement for sodium is (2,8,1). Electron arrangements for elements are given in the data booklet (see Figure 3.6).

If we examine any particular group of elements in the Periodic Table, we find that the number of electrons in the outermost energy level or shell is the same. All of the group 2 elements, for example, have two outer electrons. The same is true for group 7, the halogens – they all have seven outer electrons. The rule does not quite work for the noble gases in group 0. All of them have eight outer electrons, apart from helium which has only two.

What is very noticeable about elements in the same group is that they all have similar chemical properties. Chemists believe that this is because they have the same number of outer electrons. All of the alkali metals, for example, react with water to produce hydrogen gas. They are in group 1 of the Periodic Table and all have one outer electron (see Figure 3.7). In some versions of the Periodic Table, hydrogen is placed in group 1, but it really does not belong there.

Group 1 – the alkali metals

Li 2,1

Na 2,8,1

K   2,8,8,1

Rb 2,8,18,8,1

Cs 2,8,18,18,8,1

Fr 2,8,18,32,18,8,1

Figure 3.7 ▲
Elements in the same group have the same number of outer electrons.

Although, like the alkali metals, its atoms have one outer electron, which is also its only electron, chemically it is not like the alkali metals and is better placed away from them. In modern versions of the Periodic Table hydrogen is not assigned to any particular group.

Mass number and isotopes

The most common atom in the universe is a hydrogen atom that consists of one proton and one electron. Strangely, it possesses no neutrons, yet every other kind of atom does. There are, in fact, two other naturally occurring hydrogen atoms, one has a nucleus consisting of one proton and one neutron and the other has one proton and two neutrons in its nucleus. These different atoms of the same element can be identified by their **mass number**. Since electrons are so light in comparison to protons and neutrons, the mass of an atom is mainly determined by the total number of protons and neutrons in the nucleus. This total of protons and neutrons is called the mass number.

Mass number = number of protons + number of neutrons

For any element, the number of protons in the nucleus of its atoms is fixed and is referred to as the atomic number. The number of neutrons can vary, but can be calculated by subtracting the atomic number from the mass number.

Atomic number = number of protons

Mass number − atomic number = number of neutrons

For each different atom, a symbol can be written which shows both the mass number and the atomic number.

The symbols for the three types of hydrogen atom are as follows:

Mass number ↘ ^1_1H ^2_1H ^3_1H

Atomic number ↗

The three different types of hydrogen atom are called **isotopes**. This means that they are atoms of the same element but they have different numbers of neutrons. The isotopes of any element have the same atomic number but have different mass numbers. Almost all elements have more than one naturally occurring isotope, the number varying from two up to about ten. A few, like sodium and fluorine, have only one.

Questions

1   $^{14}_7\text{N}$ represents the most common isotope of the element nitrogen. Give the symbol for the nitrogen isotope that has 8 neutrons.

2   Give the number of protons, neutrons and electrons present in each of the following isotopes: **(a)** $^{23}_{11}\text{Na}$ **(b)** $^{35}_{17}\text{Cl}$ **(c)** $^{235}_{92}\text{U}$

Relative atomic mass

Actual masses of atoms are far too small for convenient use. It takes six million million million million hydrogen atoms, for example, to weigh just 10 g. Because of this, a relative scale is used based on one atom of carbon-12 ($^{12}_6\text{C}$) given a mass of 12 atomic mass units (amu) exactly. The effect of this is that we can treat the mass number of an atom as being a good approximation to the mass of an atom in atomic mass units. Chemists define the **relative atomic mass (RAM)** of an element as being the average of the mass numbers of its isotopes, taking into account the proportions of each. Since sodium and fluorine have only one isotope, the relative atomic mass is the mass number of that isotope in each case. The RAM of sodium is therefore 23.0 because its only isotope has a mass number of 23. Fluorine-19 is the only isotope of fluorine, which therefore has an RAM of 19.0. The relative atomic masses of most elements are not whole numbers because of the existence of isotopes. Chlorine, for example, has two isotopes, $^{35}_{17}\text{Cl}$ (75%) and $^{37}_{17}\text{Cl}$ (25%); it has an RAM of 35.5.

It is possible to calculate the RAM of any element, if we are told the mass number and percentage of each isotope, using the following formula:

$$\mathbf{RAM} = \sum \left(\frac{\% \times \mathbf{mass\ number}}{\mathbf{100}} \right)$$

$$\text{RAM of chlorine} = \frac{75 \times 35}{100} + \frac{25 \times 37}{100}$$

$$= 26.25 + 9.25$$

$$= 35.5$$

($\sum$ means 'the sum of')

Question

3 Copper consists of two isotopes with mass numbers 63 and 65. The relative atomic mass of copper is 63.5.

Which of the isotopes is present in the greater amount?

Summary

- An **atom** is the smallest part of an element that can exist.
- The **nucleus** is the extremely small central part of an atom.
- **Protons** are positively charged particles found in the nucleus.
- **Neutrons** are uncharged particles found in the nucleus.
- **Electrons** are negatively charged particles that orbit the nucleus.
- **Atomic number** = the number of protons in the nucleus.
- **Mass number** = the number of protons + neutrons in the nucleus.
- **Isotopes** are atoms with the same atomic number but different mass numbers.
- The **relative atomic mass** of an element is the average of the mass numbers of its isotopes, taking into account the proportions of each.

In questions 1–4 choose the correct word(s) **from the following list** to complete the sentence.

element, atomic number, neutrons, mass number, compound, protons, electrons

1 The number '14' in the symbol $^{14}_{7}N$ is called the _____.

2 Atoms of the same _____ all have the same atomic number.

3 The nucleus of almost all atoms is made up of neutrons and _____.

4 The numbers '2,8,1' tell us the arrangement of _____ in an atom of sodium.

5 An isotope of helium that has a mass number of 4 differs from other isotopes of helium in the number of
  A protons in the nucleus
  B electrons in the atom
  C neutrons in the nucleus
  D outer electrons

6 An element with four electrons in its outer energy level could have an atomic number of
  A 4   B 16   C 34   D 50

7 Isotopes of the same element have
  A the same number of protons and neutrons, but a different number of electrons
  B the same number of protons and electrons, but a different number of neutrons
  C the same number of neutrons, but different numbers of protons and electrons
  D the same number of protons, but different numbers of electrons and neutrons.

8 An atom is neutral because it contains
  A a number of electrons equal to the sum of the number of protons and neutrons
  B a number of neutrons equal to the sum of the number of electrons and protons
  C a number of protons equal to the number of neutrons
  D a number of electrons equal to the number of protons

9 An element that is a gas at room temperature and pressure could have the electron arrangement
  A 2, 8, 2
  B 2, 8, 4
  C 2, 8, 6
  D 2, 8, 8

10 The number of electrons in an atom is 34, and the mass number is 79. The number of neutrons in the atom is
  A 11
  B 34
  C 45
  D 79

11 Lithium and sodium are reactive metals with similar chemical properties.

  (a) Give the electron arrangements for these two elements.
  (b) What feature of their electron arrangements is responsible for lithium and sodium having similar chemical properties?

12 Use information in the data booklet to find the missing terms in the table below.

Element	Symbol	Atomic number	Electron arrangement
Lithium	Li	3	2,1
Boron	(a)	(b)	(c)
(d)	Si	(e)	(f)
(g)	(h)	20	(i)

13 The symbol for deuterium is $^{2}_{1}H$.

  (a) Complete the table to show the number of protons and neutrons in a deuterium nucleus.

	Number
Protons	
Neutrons	

  (b) What name is given to the total number of protons and neutrons in the nucleus of an atom?
  (c) What name is given to pairs of atoms such as $^{1}_{1}H$ and $^{2}_{1}H$?

14 There are two different types of lithium atom $^6_3$Li and $^7_3$Li. Lithium has a relative atomic mass of 6·9.

(a) What name is used to describe the different types of lithium atom?

(b) What can be said about the proportions of each type of atom in lithium?

15 The diagram represents the structure of an atom.

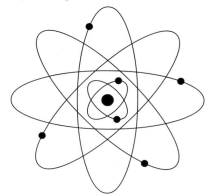

Fill in the missing information at (a), (b), (c) and (d) in the tables.

The nucleus		
Name of particle	**Relative mass**	**Charge**
Proton	1	(a)
Neutron	(b)	0

Outside the nucleus		
Name of particle	**Relative mass**	**Charge**
(c)	almost zero	(d)

16 Explain the meaning of the term 'relative atomic mass of an element' in terms of the mass numbers of the isotopes of which it is made up.

17 In a table of accurate values of relative atomic masses, magnesium is given a value of 24.3 but no magnesium isotope has this mass number. Explain.

18 Silver consists of two isotopes, ^{107}Ag and ^{109}Ag. The relative atomic mass of silver is almost exactly 108.

What does this information tell you about the relative amounts of the two isotopes present in silver?

19 With the help of information contained within the data booklet, find the missing information in the table below.

Isotope	Mass no.	Atomic no.	No. of protons	No. of neutrons	No. of electrons
$^9_4$Be	(a)	(b)	(c)	(d)	(e)
(f)	19	(g)	9	(h)	(i)
(j)	(k)	17	(l)	20	(m)

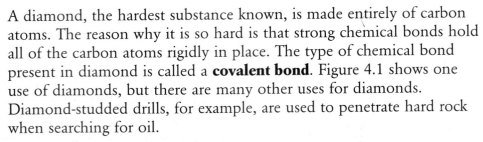

4 Bonding, Structure and Properties

Covalent bonding

A diamond, the hardest substance known, is made entirely of carbon atoms. The reason why it is so hard is that strong chemical bonds hold all of the carbon atoms rigidly in place. The type of chemical bond present in diamond is called a **covalent bond**. Figure 4.1 shows one use of diamonds, but there are many other uses for diamonds. Diamond-studded drills, for example, are used to penetrate hard rock when searching for oil.

If there were no bonds between atoms, every substance would be a gas – there could be no liquids or solids. In fact, only the noble gases exist as individual atoms, not bonded to other atoms. Chemists believe that the electron arrangements of the noble gases are particularly stable. They also believe that, when main group elements form chemical bonds they usually achieve the stable electron arrangement of a noble gas.

Covalent bonding in molecular elements

Hydrogen

One way in which atoms can achieve noble gas electron arrangements is to *share* electrons. Hydrogen atoms only have one electron, whereas the noble gas helium has two. So if two hydrogen atoms share their electrons, they each gain a share in an extra electron and achieve the same electron arrangement as helium. We call the shared pair of electrons a *single* **covalent bond**. The two hydrogen atoms which are joined by a single covalent bond are referred to as a hydrogen **molecule**, since a molecule is a group of atoms held together by covalent bonds. The formation of a hydrogen molecule from two hydrogen atoms is shown in Figure 4.2.

The single covalent bond in a hydrogen molecule can be represented by a line, so the molecule can be shown as H–H, which is called a **full structural formula**. It is more common to use the **molecular formula** for hydrogen, written as H_2, which tells us that there are two atoms in a molecule of hydrogen.

Halogens

Several non-metal elements form molecules which consist of two atoms bonded covalently. All of the halogen atoms have seven outer electrons, but the nearest noble gas atoms all have eight. In every case they form molecules by sharing an electron between two atoms and in this way achieve a noble gas electron arrangement.

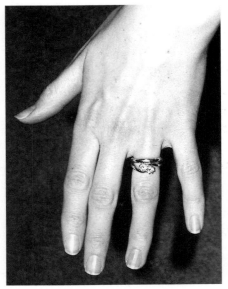

Figure 4.1 ▲
The atoms in a diamond are held together very strongly by covalent bonds.

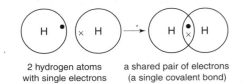

| 2 hydrogen atoms with single electrons | a shared pair of electrons (a single covalent bond) |

Figure 4.2 ▲
Two hydrogen atoms combining.

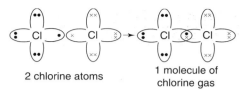

2 chlorine atoms   1 molecule of chlorine gas

Figure 4.3 ▲
Two chlorine atoms combining.

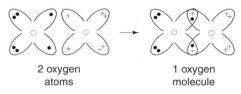

2 oxygen atoms   1 oxygen molecule

Figure 4.4 ▲
Two oxygen atoms combining.

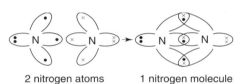

2 nitrogen atoms   1 nitrogen molecule

Figure 4.5 ▲
Two nitrogen atoms combining.

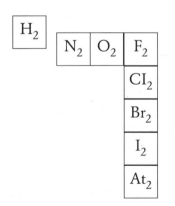

Figure 4.6 ▲
The diatomic elements.

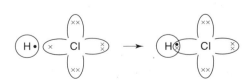

Figure 4.7 ▲
Hydrogen and chlorine atoms combining to produce a hydrogen chloride molecule.

Only outer electrons are involved in bond formation, so inner electrons need not be shown in bonding diagrams. Figure 4.3 shows how two chlorine atoms join to form a chlorine molecule by sharing electrons. Notice how, when forming covalent bonds, like most of the main group elements, the outer electrons are confined to four lobe-shaped regions called orbitals. Each lobe can only hold two electrons. A chlorine molecule can be represented by the full structural formula Cl–Cl, or by the molecular formula Cl_2.

Oxygen

Oxygen atoms only have six outer electrons, which is two less than the atoms of the nearest noble gas neon. When forming an oxygen molecule, two oxygen atoms achieve a noble gas electron arrangement by sharing two pairs of electrons, meaning that the atoms are held together by a *double* covalent bond (see Figure 4.4). Notice that in this case the six outer electrons occupy the four lobes so that two are full and two are half full. In general, when adding electrons to lobes, they are added one-by-one to each lobe before putting two side by side in the same lobe.

An oxygen molecule can be represented by the full structural formula O=O, where '=' represents a double covalent bond, or by O_2 as the molecular formula.

Nitrogen

Nitrogen atoms have only five outer electrons, which means that each atom needs to obtain a share in another three electrons. Each nitrogen atom has one filled lobe and three half-filled lobes. Figure 4.5 shows how two nitrogen atoms form a nitrogen molecule by sharing three pairs of electrons to make a *triple* covalent bond.

A nitrogen molecule is therefore represented by the full structural formula N≡N, or by the molecular formula N_2.

Diatomic elements

Hydrogen, nitrogen, oxygen and the halogens are said to be **diatomic** because their molecules consist of two atoms bonded together (see Figure 4.6).

Covalent bonding in molecular compounds

The atoms of *different* non-metal elements can also bond covalently, with main group elements achieving noble gas electron arrangements by sharing electrons.

Hydrogen chloride

The single electron from a hydrogen atom is shared with an electron from a chlorine atom, forming a single covalent bond. Hydrogen achieves the electron arrangement of helium and chlorine that of argon (see Figure 4.7).

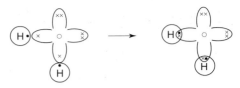

Figure 4.8 ▲
Hydrogen and oxygen atoms combining to produce a water molecule.

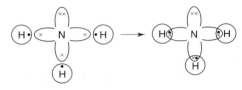

Figure 4.9 ▲
Hydrogen and nitrogen atoms combining to produce an ammonia molecule.

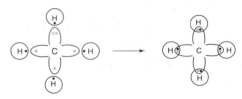

Figure 4.10 ▲
Hydrogen and carbon atoms combining to produce a methane molecule.

Hydrogen chloride exists as diatomic molecules with a full structural formula of H–Cl and a molecular formula of HCl.

Water (hydrogen oxide)

Oxygen atoms have six outer electrons and therefore need two more electrons in order to achieve a noble gas electron arrangement. Since each hydrogen atom can only supply one electron, it takes two hydrogen atoms to bond covalently with one oxygen atom (see Figure 4.8). The molecular formula of water is H_2O and the full structural formula is:

$$\begin{array}{c} O \\ H \quad\quad H \end{array}$$

Ammonia (nitrogen hydride)

Nitrogen atoms have five outer electrons, so three lobes have only one electron. Each nitrogen atom needs three hydrogen atoms to provide the extra three electrons, then, by sharing, both nitrogen and hydrogen achieve noble gas electron arrangements (see Figure 4.9). The compound formed is called ammonia, with molecular formula NH_3. The full structural formula is:

$$\begin{array}{c} H\!-\!N\!-\!H \\ | \\ H \end{array}$$

Methane (carbon hydride)

Carbon atoms have four outer electrons and therefore need four hydrogen atoms to supply the extra four electrons required to achieve the same electron arrangement as neon (see Figure 4.10). The compound produced is called methane and its molecular formula is CH_4. The full structural formula is:

$$\begin{array}{c} H \\ | \\ H\!-\!C\!-\!H \\ | \\ H \end{array}$$

Simpler electron diagrams

Electron diagrams for both atoms and molecules can be simplified by leaving out the boundaries of the orbitals. This allows more compact diagrams to be drawn as shown in the table below.

Hydrogen	Chlorine	Oxygen	Nitrogen
H ⦁ₓ H	⦂ Cl ⦂ Cl ×ₓ	⦁ O ⦁ O ×ₓ	⦁ N ⦂ N ×ₓ
Hydrogen chloride	**Water (hydrogen oxide)**	**Ammonia**	**Methane**
H ⦂ Cl ⦂	H ⦂ O ⦂ H	H ⦂ N ⦂ H H	H ⦂ C ⦂ H H H

Table 4.1 ▲

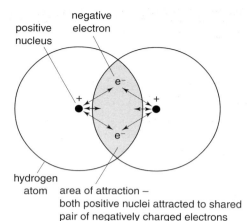

positive nucleus

negative electron

e⁻

+ +

e⁻

hydrogen atom

area of attraction – both positive nuclei attracted to shared pair of negatively charged electrons

Figure 4.11 ▲
Two hydrogen atoms sharing a pair of electrons.

Question

1 Draw diagrams showing outer electrons to indicate how molecules of the following substances form from their atoms:

(a) bromine, Br_2; (b) hydrogen fluoride, HF; (c) silicon hydride, SiH_4; (d) phosphorus hydride, PH_3; (e) hydrogen sulphide, H_2S; (f) tetrachloromethane, CCl_4.

What is a covalent bond?

Why do atoms stay joined together when they share a pair of electrons? The answer lies in the considerable attraction between the negatively charged, shared electrons and the positively charged nuclei. Consider the hydrogen molecule that is shown in Figure 4.11. The attraction between shared electrons and nuclei pulls the atoms closer together until a 'position of balance' is reached when the repulsion between the positively charged nuclei is balanced by the attraction of the nuclei for the shared electrons. This is why the two hydrogen atoms stay bonded together, as do all atoms which are joined by covalent bonds.

Chemists define a **covalent bond** as two positive nuclei held together by their common attraction for a shared pair of electrons.

The true shapes of molecules

The diagrams of molecules that have been used so far have been simplifications; they do not show their true shapes, except in the simplest cases. Fortunately, there is an easy way of predicting the shape of a molecule which is based on assuming that pairs of outer electrons, whether bonded or not, will repel other pairs. This often leads to a basically tetrahedral shape, as the lobes in which they are present point towards the corners of a tetrahedron (see Figure 4.12).

When drawing the true shape of a molecule, which is called a **perspective formula**, special conventions are used. A solid line shows a bond in the same plane as the paper; a dotted line, a bond behind the paper; and a wedge shape is used to show a bond in front of the paper. Table 4.2 shows formulae representing four compounds each containing two elements.

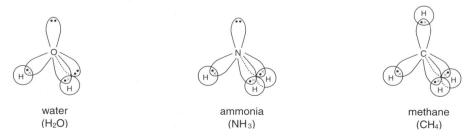

water
(H₂O)

ammonia
(NH₃)

methane
(CH₄)

Figure 4.12 ▲
The shapes of these molecules are based on tetrahedra.

Compound	Molecular formula	Full structural formula	True shape (Perspective formula)
Hydrogen chloride	HCl	H—Cl	H—Cl (linear)
Water	H_2O	H—O 　| 　H	H—O 　　＼ 　　　H (bent or angular)
Ammonia	NH_3	H—N—H 　　| 　　H	(pyramidal)
Methane	CH_4	H 　　| H—C—H 　　| 　　H	(tetrahedral)

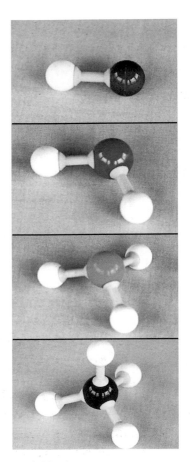

Figure 4.13 ▲
Molecular models of hydrogen chloride, water, ammonia and methane.

Table 4.2 ▲

Question

2 Draw perspective formulae (those that show the true shape of the molecules) for:

(a) hydrogen fluoride, HF; (b) hydrogen sulphide, H_2S;
(c) phosphorus hydride, PH_3; (d) silicon hydride, SiH_4;
(e) tetrachloromethane, CCl_4; (f) dichloromethane, CH_2Cl_2.

Polar covalent bonding

When atoms of the same element are bonded covalently, the electrons in the bond are shared equally by both atoms and the bond is said to be non-polar. However, atoms of different elements can have different attractions for the shared electrons in a covalent bond. The atom with the greater attraction for electrons acquires a slight negative charge ($\delta-$), while the other atom becomes slightly positively charged ($\delta+$).

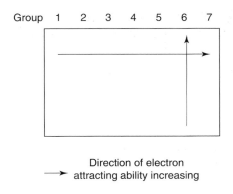

Group 1 2 3 4 5 6 7

→ Direction of electron attracting ability increasing

Figure 4.14 ▲
Changes in electron attracting ability.

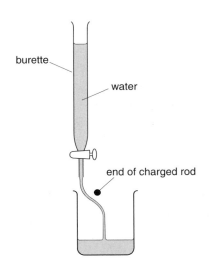

burette

water

end of charged rod

Figure 4.16 ▲
A thin stream of water is attracted to a charged plastic rod.

Covalent bonds with unequal sharing of electrons are referred to as **polar covalent bonds**.

For the main groups 1–7 in the Periodic Table, ignoring the noble gases, there are clear trends in the electron attracting abilities of the atoms of the different elements. These are shown in Figure 4.14.

Slight differences in electron attracting ability have no great effect on the properties of a compound, but larger differences can. Of particular interest are compounds in which hydrogen atoms are bonded to oxygen, nitrogen, chlorine or fluorine atoms, as in all cases the covalent bonds are very polar (see Figure 4.15).

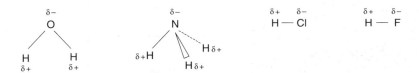

Figure 4.15 ▲
Four molecules that contain polar covalent bonds.

The presence of polar covalent bonds in a compound can cause it to have unusual properties. A thin stream of water, for example, is attracted to a charged plastic rod (see Figure 4.16). If *any* liquid behaves like this, then it contains polar covalent bonds. Water also has unusually high freezing and boiling points because the polar ends of the molecules attract neighbouring molecules, making them more difficult to separate as the solid melts, or as the liquid boils. This is also true of ammonia, hydrogen chloride and hydrogen fluoride.

Questions

3 Given that sulphur atoms have a greater attraction for bonded electrons than hydrogen atoms, draw a formula for a molecule of hydrogen sulphide that shows both its shape and the polarity of its bonds.

4 Use the trends in electron-attracting ability described earlier to help you show the polarity of the bonds present in a molecule of tetrafluoromethane, CF_4.

Properties of covalent molecular substances

Although the covalent bonds within molecules are strong, the forces of attraction between discrete molecules are weak. As a result, discrete molecular substances are either gases, liquids or low melting point solids. Between all molecules in the solid or liquid state there are weak forces of attraction called **van der Waals' forces**, which become larger as molecular size increases.

43

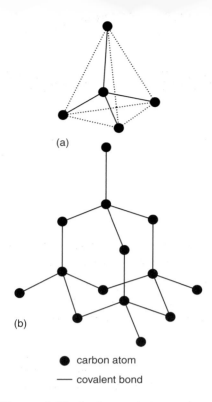

(a)

(b)

● carbon atom

— covalent bond

Figure 4.17 ▲

Structure of diamond (a) basic tetrahedral unit, (b) the diamond network structure.

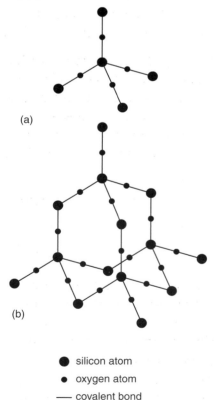

(a)

(b)

● silicon atom

• oxygen atom

— covalent bond

Figure 4.18 ▲

Structure of silicon dioxide (a) basic unit, (b) the silicon dioxide network structure.

Some molecules which contain polar covalent bonds have additional forces of attraction between them, but these are still weak in comparison with the strength of covalent bonds. Because molecules are, overall, electrically neutral, covalent molecular substances do not conduct electricity in any of the three states of matter – solid, liquid or gas.

Covalent network substances

Diamonds, which were referred to earlier, contain only carbon atoms linked by covalent bonds, but whereas many other non-metal elements have low melting points, diamond has a very high melting point. Since it is also the hardest substance known, it is obviously different from covalent molecular elements like oxygen, O_2, or iodine, I_2.

Diamond is an example of a **covalent network** element. All of the carbon atoms in a diamond are held in a tetrahedral arrangement linked by single covalent bonds (see Figure 4.17). The covalent network structure of a diamond is very strong and rigid, two features which account for its great hardness. It should be noted that even a small diamond will contain many millions of carbon atoms, so the structure shown in Figure 4.17 is only the tiniest part of a diamond.

Graphite is another form of carbon that exists as a covalent network. Graphite has a layered structure, with only weak van der Waals' forces between the layers. As a result, although the melting point is high, graphite is weak and brittle whereas diamond is strong and hard. Another important difference is that graphite conducts electricity, whereas diamond does not. Within the graphite structure some electrons are free to move anywhere in a given layer – they are said to be delocalised.

Some compounds exist as covalent networks. For example, silicon dioxide has a structure similar to that of diamond, with millions of silicon and oxygen atoms held rigidly by covalent bonds (see Figure 4.18). The melting point is high, about 1700°C, and silicon dioxide does not conduct electricity when molten. The reason for the high melting point is that a high temperature is needed to break the strong covalent bonds. Silicon dioxide is found widely in nature, as sand and quartz for example. The chemical formula for silicon dioxide is SiO_2, which is an **empirical formula**, not a molecular formula, since it only shows the ratio of atoms present in the covalent network structure.

Silicon carbide, SiC, is a covalent network compound that is almost as hard as diamond. The structure is the same as that of diamond, with half of the carbon atoms replaced by silicon atoms. SiC is also an empirical formula, indicating the ratio of atoms present in the network. Silicon carbide is also known by the name carborundum and can be found in DIY shops as sharpening stones for tools and in the form of dark grey abrasive paper.

Summary

Type of covalent substance	Molecular	Network
Melting point	low	high
Strength of bonds broken	weak	strong
Type of bond broken	van der Waals'	covalent

Ionic bonding

In covalent bonding, main group elements achieve noble gas electron arrangements by *sharing* electrons. Another way of achieving the electron arrangement of a noble gas is by atoms *losing* or *gaining* electrons. This is what happens in ionic bonding, resulting in the formation of **ionic bonds**. Compounds formed from metals and non-metals usually contain ionic bonds.

Sodium chloride (common salt) contains ionic bonds. Sodium atoms have an electron arrangement of 2,8,1. The nearest noble gas, neon, has an electron arrangement of 2,8. Sodium atoms achieve this arrangement by losing one outer electron. Chlorine atoms have an electron arrangement of 2,8,7, so they can accept the electron lost by sodium, making their electron arrangement 2,8,8. Chlorine atoms achieve the electron arrangement of the noble gas argon by doing this.

When atoms lose or gain electrons, electrically charged particles called **ions** are formed. Atoms are electrically neutral, so gaining an electron results in an ion with a single negative charge. The chlorine atom gains an electron and forms a chlor*ide* ion, written as Cl^-. Loss of an electron by a sodium atom results in the formation of a sodium ion with a single positive charge, written as Na^+. The formation of sodium chloride can be summed up as follows:

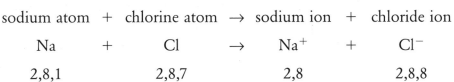

sodium atom + chlorine atom → sodium ion + chloride ion

$$Na \quad + \quad Cl \quad \rightarrow \quad Na^+ \quad + \quad Cl^-$$

2,8,1　　　　　2,8,7　　　　　2,8　　　　　2,8,8

Magnesium oxide is also ionic. Magnesium atoms have an electron arrangement of 2,8,2, which is two more than the nearest noble gas neon. Oxygen atoms have electron arrangement 2,6, so they can accept the two electrons from a magnesium atom to give them the same noble gas arrangement. The loss of two electrons by a magnesium atom produces a magnesium ion with two positive charges, written Mg^{2+}. The gain of two electrons by an oxygen atom produces an ox*ide* ion with two negative charges, written O^{2-}.

Figure 4.19 ▲
Sodium chloride (common salt) is an ionic compound.

The formation of magnesium oxide can therefore be written as:

magnesium atom + oxygen atom → magnesium ion + oxide ion

$$\text{Mg} \qquad + \qquad \text{O} \qquad \rightarrow \qquad \text{Mg}^{2+} \qquad + \qquad \text{O}^{2-}$$

$$2,8,2 \qquad\qquad\qquad 2,6 \qquad\qquad\qquad 2,8 \qquad\qquad\qquad 2,8$$

The charge that an ion of a main group element will have depends on which group it belongs to in the Periodic Table. Metals in groups 1, 2 and 3 lose their outer electrons, while non-metals in groups 5, 6 and 7 gain electrons to bring the number up to 8. By doing this both metals and non-metals achieve noble gas electron arrangements. Note that the elements in group 4 do not usually form ions.

Question

5 Write formulae for the ions formed by:
   (a) calcium, **(b)** lithium, **(c)** fluorine, **(d)** nitrogen, **(e)** potassium.

Group number	1	2	3	4	5	6	7
Charge on ion	1+	2+	3+	/	3−	2−	1−
Example of ion	Na^{+}	Mg^{2+}	Al^{3+}	/	P^{3-}	S^{2-}	Cl^{-}

Ionic lattices

So far we have considered individual ions, but the force of electrostatic attraction between oppositely charged ions, called the ionic bond, is not directional, like the covalent bond, so the ions cluster together. The positive ions surround the negative ions and the negative ions surround the positive ions. In this way a huge well-ordered structure is built up called an **ionic lattice**. Sodium chloride crystals are cubic because the ions form a cubic lattice (see Figure 4.20). Within the sodium chloride lattice, each chloride ion is surrounded by six sodium ions and each sodium ion by six chloride ions.

Within the lattice the ions are held tightly and can only vibrate about fixed positions. As a result they do not conduct electricity in the solid state because the ions, although electrically charged, are not free to move about. The melting points of ionic compounds are high because ionic bonds are strong. Sodium chloride, for example, melts at just over 800°C. When molten, the ions are free to move and ionic compounds then conduct electricity. Many of them dissolve in water, which also frees the ions from the lattice and, as a result, aqueous solutions of ionic compounds conduct electricity as well.

Electrolysis

The word **electrolysis** is used to describe the flow of ions through solutions or molten compounds with chemical changes taking place at the electrodes. Note that the literal meaning of 'electrolysis' is 'breaking up by means of electricity'.

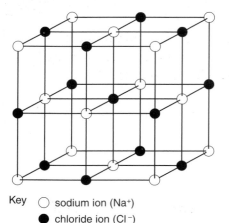

Key ○ sodium ion (Na^{+})
     ● chloride ion (Cl^{-})

Figure 4.20 ▲
The sodium chloride lattice.

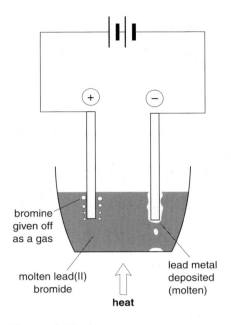

Figure 4.21 ▲
Electrolysis of molten lead(II) bromide.

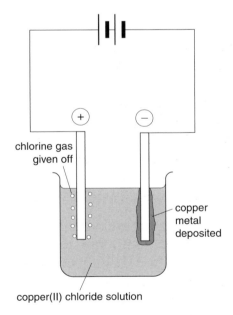

Figure 4.22 ▲
Electrolysis of copper(II) chloride solution.

When a direct current is passed through a molten ionic compound, the positively charged ions are attracted to the negative electrode and the negatively charged ions to the positive electrode. A substance that conducts due the movement of ions is called an **electrolyte.** Lead(II) bromide is often used to show the flow of electricity through a molten ionic compound because it has a fairly low melting point. (Lead(II) tells us that the lead ion present has two positive charges and is written as Pb^{2+}). Chemical changes are seen to take place at the electrodes, with silvery molten lead forming at the negative electrode and reddish-brown bromine gas at the positive electrode (see Figure 4.21).

The current of electricity breaks up the molten lead(II) bromide into lead and bromine. Electrolysis of any molten ionic compound, made up of positive metal ions and negative non-metal ions, results in the metal being formed at the negative electrode and the non-metal at the positive electrode. Positive metal ions gain electrons at the negative electrode and negative non-metal ions lose electrons at the positive electrode. During the electrolysis of molten lead(II) bromide the **ion-electron equations** for the reactions taking place at the electrodes are as follows:

At the negative electrode: $Pb^{2+} + 2e^- \rightarrow Pb$

At the positive electrode: $2Br^- \rightarrow Br_2 + 2e^-$

Electrolysis of aqueous solutions of certain ionic compounds can also make them break up. During the electrolysis of copper(II) chloride solution, for example, copper is formed at the negative electrode and chlorine gas is formed at the positive electrode (see Figure 4.22). The ion-electron equations for the reactions taking place are:

At the negative electrode: $Cu^{2+} + 2e^- \rightarrow Cu$

At the positive electrode: $2Cl^- \rightarrow Cl_2 + 2e^-$

Question

6 Give ion-electron equations for the reactions taking place at each electrode during the electrolysis of: **(a)** molten nickel(II) fluoride, **(b)** molten potassium iodide, **(c)** zinc(II) chloride solution.
Note that in all three cases the electrode products are a metal element and a halogen.

Direct current and electrolysis

All electrolysis experiments use direct current and not alternating current. If a direct current is used then one electrode remains positive and the other negative.

If an alternating current is used then the electrical sign of each electrode is constantly changing between positive and negative. This means that both products are formed at each electrode and the individual electrode products cannot be identified. Using direct current, the products are formed at different electrodes, making it possible to separate and identify these products.

 ## Prescribed Practical Activity

Safety note

Because of the toxic nature of chlorine, only a tiny amount should be sniffed (by the 'wafting' technique).

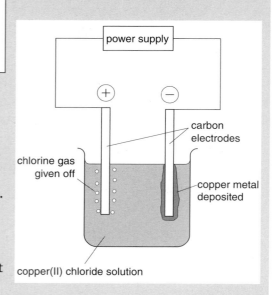

Electrolysis

- The aim of this experiment is to electrolyse copper(II) chloride solution and to identify the products at the positive and negative electrodes.
- Copper is formed as a brown solid on the negative electrode.
- Chlorine gas, which is poisonous and has a choking smell, is formed at the positive electrode.
- A piece of moist pH paper, or moist blue litmus paper, held above the positive electrode first of all turns red to show that chlorine produces an acidic solution, and is then bleached.

Ions containing more than one kind of atom

So far we have looked at ions consisting of only one kind of atom, like the sodium ion, Na^+, and the oxide ion, O^{2-}. However, there are also ions with two or more different kinds of atom. The sulphate ion, for example, contains sulphur and oxygen. Its chemical formula is SO_4^{2-}. This means that it is a group of five atoms, made up of one sulphur atom and four oxygen atoms, with two negative charges. This and some other **group ions** that you are likely to meet are given in Table 4.3 below.

one positive	one negative	two negative	three negative
ammonium, NH_4^+	hydroxide, OH^- nitrate, NO_3^- permanganate, MnO_4^-	carbonate, CO_3^{2-} chromate, CrO_4^{2-} dichromate, $Cr_2O_7^{2-}$ sulphate, SO_4^{2-}	phosphate, PO_4^{3-}

Table 4.3 ▲

A more comprehensive table of ions containing more than one kind of atom is given in the data booklet.

Investigating the colour of ions

Many ionic compounds are colourless. When their crystals are crushed they produce white powders and when they dissolve they produce colourless solutions. We can therefore conclude that both of the ions present are colourless. For example, sodium chloride is colourless, so we know that both sodium ions and chloride ions are colourless.

Sodium chromate, however, is an attractive yellow colour. The colour must be due to the chromate ion because, as we know by comparison with sodium chloride, sodium ions are colourless.

It is mostly the presence of transition metals in ionic compounds that seem to be the cause of colour. Copper(II) sulphate is blue, whereas potassium sulphate is colourless. Evidently potassium and sulphate ions are colourless and the colour of copper(II) sulphate is due to the blue copper(II) ion.

Table 4.4 contains information about some coloured ionic compounds.

Questions

7 Iron(II) sulphate is green, but sodium sulphate is colourless. Write the name and formula of the ion which is responsible for the green colour.

8 Potassium permanganate is purple, but potassium sulphate is colourless. Write the name and formula of the ion which is responsible for the purple colour.

9 Use the information in Table 4.4 to suggest the colour of a solution of copper(II) chromate.

Compound	Colour	Ion responsible for the colour
copper(II) sulphate copper(II) nitrate	blue	copper(II), Cu^{2+}
nickel(II) sulphate nickel(II) chloride	green	nickel(II), Ni^{2+}
sodium chromate potassium chromate	yellow	chromate, CrO_4^{2-}
ammonium dichromate potassium dichromate	orange	dichromate, $Cr_2O_7^{2-}$

Table 4.4 ▲
Colours of some ionic compounds.

Ion migration experiments

Positive ions in solution are attracted to negative electrodes and negative ions in solution are attracted to positive electrodes. We can use this information to construct experiments in which the movement of ions can be seen as different colours moving towards electrodes. These are called *ion migration experiments*.

In Figure 4.23, a direct current is passed through a solution of copper(II) chromate. Slowly a blue colour moves through the colourless electrolyte towards the negative electrode. At the same time a yellow colour moves towards the positive electrode.

Figure 4.23 ▲
Ion migration experiment using a U-tube and copper(II) chromate solution.

49

Question

10 An ion migration experiment was set up as shown in Figure 4.24. From the first salt (A), a green colour moved towards the negative electrode. From the second salt (B), an orange colour moved towards the positive electrode. The third salt (C) was potassium permanganate, from which a purple colour moved towards the positive electrode. Give the sign of the charges carried by each of the coloured ions present and suggest a possible name and formula for each.

We already know that copper(II) ions are blue and that chromate ions are yellow. This experiment enables us to learn two more things about these ions. The first is that copper(II) ions have a positive charge because they are attracted towards a negative electrode. The second is that chromate ions must have a negative charge because they are attracted to a positive electrode.

A second type of ion migration experiment uses filter paper soaked in a colourless salt solution such as potassium nitrate. This is placed on a glass or plastic surface. Bands of colour are seen to move slowly across the filter paper.

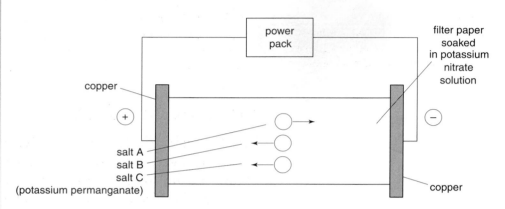

Figure 4.24 ▲
Ion migration experiment using filter paper soaked in potassium nitrate solution.

Bonding and solubility

Ionic compounds are usually soluble in water, whereas covalent compounds are usually not. Most data books contain information about the solubility of ionic compounds and many are described as being 'very soluble'. This applies, for example, to nitrates, all of which are very soluble in water. In the case of sulphates, most are very water-soluble, but some are less soluble, with barium sulphate being one of the most insoluble of the ionic compounds.

When an ionic compound dissolves in water the lattice of ions breaks down. This happens because the water molecules, due to the special nature of their polar covalent bonds, are attracted to ions in the lattice and can pull them from their lattice positions.

Water molecules form only weak forces of attraction with atoms in covalent network compounds and, as a result, cannot break the covalent bonds holding the network together. Sand (silicon dioxide) therefore does not dissolve in water.

The only covalent molecular compounds that dissolve readily in water are those which, like water, contain a special type of polar covalent bond so that their molecules can be attracted to the water molecules. This is true of sugars and simple alcohols.

Figure 4.25 ▲
The iodine here is dissolved in ethanol. Several non-aqueous solvents are used to make nail varnish and its remover.

Covalent molecular substances are often soluble in non-aqueous solvents, whereas ionic compounds are not. The covalent molecular substances candle wax and iodine, for example, dissolve in hexane, whereas the ionic compounds sodium chloride and copper sulphate do not. Non-aqueous solvent molecules, like hexane, do not attract the ions strongly enough to pull them from their lattice positions.

The results of two solubility experiments are shown in Table 4.5.

Solute	Solvent	
	Water	**Hexane**
candle wax (covalent)	insoluble	soluble
sodium chloride (ionic)	soluble	insoluble

Table 4.5 ▲

Metallic bonding

Metals consist of a lattice of positively charged ions held together by their electrostatic attraction for delocalised electrons from their outermost shell (energy level). This attraction is called **metallic bonding**. The electrons from the outer shell are free to move throughout the structure of the metal and are responsible for the conduction of electricity by the metal as shown in Figure 4.26.

Metallic bonding has some of the characteristics of covalent bonding, in that electrons are shared, and some characteristics of ionic bonding because ions are present. Like both covalent bonding and ionic bonding, metallic bonding is usually strong, resulting in most metal elements having high melting points.

The ions in a pure metal are all of the same size and because of this they can move from one point in the lattice to the next without breaking the metal apart. Metals can therefore bend without breaking and can be beaten or rolled into thin sheets. They are said to be **malleable**.

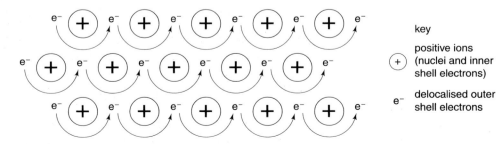

key

positive ions (nuclei and inner shell electrons)

e⁻ delocalised outer shell electrons

Figure 4.26 ▲
Conduction of electricity by a metal.

A summary of electrical conduction

Two kinds of particle can carry an electric current through a material. These are delocalised electrons, or ions that are free to move. Delocalised electrons are found in metals and in graphite. These are the only solids that conduct electricity and they do so with no chemical changes taking place. Ionic compounds do not conduct electricity in the solid state because their ions are not free to move. They do, however, conduct when molten or when dissolved in water because the ions are then free to move and can carry the current. When ionic compounds conduct, chemical reactions take place at the electrodes. Molten ionic compounds and solutions of ionic compounds are called electrolytes.

Covalent substances, apart from graphite, do not conduct electricity in the solid or liquid state. Most are insoluble in water, or form non-conducting solutions. Some covalent compounds are broken up into ions on dissolving in water (see Section 3).

Type of Substance	Does conduction take place?		
	Solid state	Liquid state	Solution in water
Ionic	no	yes	yes
Covalent	no	no	no (not usually)
Metallic	yes	yes	insoluble

Table 4.5 ▲

Substances can be classified as follows, based on their ability to conduct electricity:

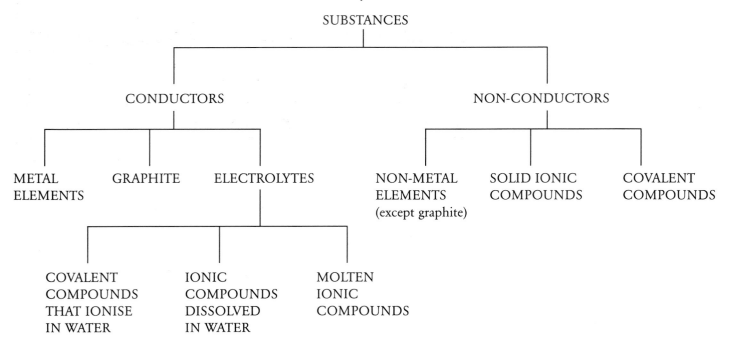

Apparatus used to test for electrical conduction

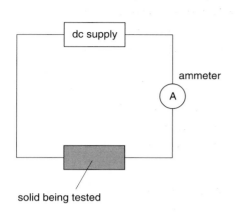

Figure 4.27 ▲
Testing a solid for electrical conduction.

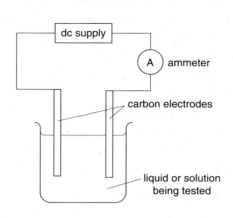

Figure 4.28 ▲
Testing a liquid or solution for electrical conduction.

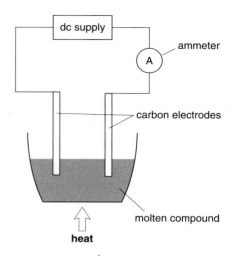

Figure 4.29 ▲
Testing a molten compound for electrical conduction.

Questions

11 A compound dissolves in water but not in hexane. A solution of the compound in water conducts electricity.

What type of bonding is likely to be present in this compound?

12 Complete the table by adding 'conducts' or 'does not conduct' in the spaces.

Substance	Solid state	Liquid state
Sulphur	_____	_____
Iron	conducts	_____
Calcium oxide	_____	_____

In questions 1–4 choose the correct word(s) **from the following list** to complete the sentence.

molecular, pyramidal, positively, sulphur, negatively, oxygen, network, tetrahedral

1 An example of an element which exists as a diatomic molecule is _____.

2 Methane molecules have a _____ shape.

3 Atoms of metals from group 1 each lose one electron to form _____ charged ions.

4 Covalent _____ substances have high melting points.

5 Which of the following does **not** have the same number of electrons as a magnesium ion?
   A A fluoride ion
   B A sodium atom
   C A neon atom
   D A sodium ion

6 Which of the following pairs of ions both have the same electron arrangement as the noble gas argon?
   A K^+ and F^-
   B Na^+ and I^-
   C K^+ and Cl^-
   D Li^+ and Br^-

7 The table below shows the ability of substances to conduct electricity.

	Solid	Liquid	Solution in water
A	No	No	No
B	No	No	Yes
C	No	Yes	Yes
D	Yes	Yes	Insoluble

Which of the above substances, A, B, C or D, could be calcium chloride?

8 Copper is a good conductor of electricity because
   A the atoms are free to vibrate
   B the atoms are in close contact
   C the atoms have electron arrangement 2,8,18,1
   D electrons can move readily from one atom to the next.

9 Dry ice (solid carbon dioxide) is a non-conductor of electricity because
   A it is a covalent compound
   B its ions are not free to move
   C it contains too few ions to carry the current
   D it contains no electrons.

10 Which of the following is the most likely shape of a molecule of silicon chloride?

   A   Cl——Si——Cl

   B   Si
           Cl   Cl
               Cl

   C   Cl
                   |
               Si
           Cl   Cl
               Cl

   D   Cl
               |
       Cl——Si——Cl
               |
               Cl

11 Which of the following elements exist as diatomic molecules?
   A Sulphur
   B Argon
   C Aluminium
   D Bromine

12 Which of the following elements exist as single atoms not bonded to other atoms?
   A Helium
   B Hydrogen
   C Sodium
   D Magnesium

13 Tetrafluoromethane is a covalent compound. Its formula is CF_4.

   (a) Draw a diagram to show the **shape** of a molecule of tetrafluoromethane.

   (b) The atoms in a hydrogen molecule are held together by a covalent bond.

A covalent bond is a shared pair of electrons.

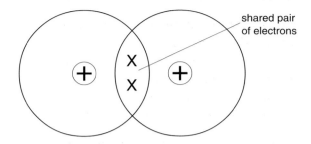

shared pair of electrons

Explain how this holds the atoms together.

14 Hydrazine is a liquid at room temperature and has a chemical formula of N_2H_4.

What type of bonding is likely to be present in hydrazine?

15 Silicon dioxide, SiO_2, and silicon fluoride, SiF_4, do not conduct electricity in the liquid state. However, silicon dioxide is a solid with a melting point of $1610^{\circ}C$, whereas silicon fluoride is a gas at room temperature.

(a) How do the structures of these two compounds differ?
(b) Why is there such a large difference in melting points? (Make reference to the bonds broken at the melting point.)

16 Carbon dioxide is a linear molecule that contains two polar covalent bonds because oxygen atoms attract bonded electrons more strongly than carbon atoms.

Draw a full structural formula for carbon dioxide showing the polar covalent bonds using the charge symbols $\delta+$ and $\delta-$.

17 Nickel carbonyl, $Ni(CO)_4$, is a liquid with a boiling point of $43^{\circ}C$.
Give as much information as you can regarding the likely bonding, structure and ability to conduct electricity of this compound.

18 Draw diagrams showing outer electrons to indicate the bonding present in molecules of:

(a) fluorine; (b) oxygen; (c) water;
(d) germanium chloride, $GeCl_4$;
(e) trichloromethane, $CHCl_3$.

19 Write formulae for the ions formed by:

(a) strontium; (b) bromine; (c) aluminium;
(d) rubidium; (e) phosphorus; (f) oxygen.

20 The table contains information about the colours of aqueous solutions of several ionic compounds.

Compound	Solution colour
Potassium chloride	colourless
Sodium nitrate	colourless
Potassium chromate	yellow
Lead nitrate	colourless
Sodium dichromate	orange
Ammonium chloride	colourless
Ammonium dichromate	orange

What is the colour of the following ions?
(a) nitrate; (b) dichromate; (c) ammonium;
(d) chromate; (e) lead; (f) sodium.

21 A compound has a melting point of $373^{\circ}C$ and is slightly soluble in water at $25^{\circ}C$. The solid compound does not conduct electricity, but it does when molten with chemical reactions taking place at the electrodes.

What type of bonding is present in the compound?

22 Calcium chloride does not conduct in the solid state, but does when molten or in aqueous solution. Explain why this is so.

23 (a) Explain why sodium chloride has a high melting point.
(b) Copper(II) chloride was electrolysed in the apparatus shown:

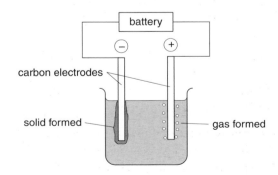

Write ion-electron equations for the formation of:
(i) the solid;
(ii) the gas.

5 Chemical Symbolism

Simple formulae for two-element compounds

In the last section molecular formulae for substances like water, H_2O, were worked out using electron arrangements for the atoms involved. There is, however, a way of predicting the *simple* chemical formula for a two-element compound using the system of **valency numbers**. Simple chemical formulae for compounds, both covalent and ionic, indicate the ratio of atoms present. In some cases, for example water, H_2O, the simple chemical formula is also the molecular formula.

The valency number for an element is given by the size of the charge on its ion, or the number of single covalent bonds that one atom of it can form. Thus calcium, which forms the ion Ca^{2+} has a valency of 2 and can therefore combine with two chloride ions, Cl^-, which have a valency of 1. This gives a simple chemical formula for the compound calcium chloride of $CaCl_2$.

Similarly, oxygen atoms that can form two single covalent bonds $-O-$ and have a valency of 2, can combine with two hydrogen atoms which can form only one single covalent bond $H-$ and therefore have a valency of 1. This gives a simple chemical formula for water (hydrogen oxide) of H_2O.

The valency numbers for the main group elements in the Periodic Table are as follows:

Group number	1	2	3	4	5	6	7	0
Valency	1	2	3	4	3	2	1	0

By following a step-by-step process, the simple chemical formula for any compound formed between two main group elements can be found.

Example 1
Work out the simple formula for the compound formed between silicon and oxygen.

Step 1 symbols   Si   O
Step 2 valencies   4   2 (silicon is in group 4
                                                             oxygen is in group 6)

Step 3 cross over the valencies Si_2O_4
Step 4 cancel out any common factor Si_1O_2
Step 5 omit '1' if present SiO_2

The simple chemical formula for the compound formed between silicon and oxygen is SiO_2. (If you look back at page 44 you will find that this is indeed the correct chemical formula for the covalent network compound silicon dioxide.)

Example 2
Work out the simple chemical formula for calcium chloride.

Step 1	symbols	Ca Cl
Step 2	valencies	2 1
Step 3	cross over valencies	Ca_1Cl_2
Step 4	cancel out any common factor (not required in this case)	Ca_1Cl_2
Step 5	omit '1' if present	$CaCl_2$

The simple chemical formula for calcium chloride is $CaCl_2$.

Questions

1 Write simple chemical formulae for:

(a) boron fluoride
(b) carbon sulphide
(c) calcium iodide
(d) hydrogen fluoride
(e) sodium sulphide
(f) lithium nitride
(g) tin bromide
(h) strontium bromide
(i) aluminium oxide
(j) magnesium phosphide
(k) hydrogen sulphide
(l) calcium nitride
(m) carbon chloride
(n) barium oxide
(o) potassium oxide
(p) calcium hydride

2 Not every compound has a chemical formula that can be predicted by the valency number method.

Which of the following formulae for oxides of nitrogen could be predicted using this method?

N_2O; NO; N_2O_3; NO_2; N_2O_5.

Prefix	Meaning
mono-	one
di-	two
tri-	three
tetra-	four
penta-	five
hexa-	six
hepta-	seven

Table 5.1 ▲

Compound formulae from prefixes

The valency number method does not work for every compound and sometimes it is not needed. Fortunately, in some cases, the compound's name helps in the working out of the chemical formula. The information is contained in the prefixes 'mono-', 'di-', 'tri-', etc. If only one prefix is given, then it may be assumed that only one atom of the other element is present in the formula. Some common prefixes and their meanings are shown in Table 5.1.

Example 1

Carbon monoxide (mono- = 1)
The simple chemical formula is CO.

Example 2

Carbon dioxide (di- = 2)
The simple chemical formula is CO_2.

Example 3

Dinitrogen monoxide
The simple chemical formula is N_2O.

Example 4

Phosphorus trichloride (tri- = 3)
The simple chemical formula is PCl_3.

Question

3 Write simple chemical formulae for:

(a) nitrogen monoxide
(b) diphosphorus trioxide
(c) silicon tetrachloride
(d) boron trifluoride
(e) divanadium pentoxide

(f) tetraarsenic hexaoxide
(g) nitrogen dioxide
(h) disulphur dichloride
(i) uranium hexafluoride
(j) dichlorine heptaoxide

Simple formulae for compounds containing group ions

We met **group ions** in the previous section on page 48. They are ions that contain more than one kind of atom such as the ammonium ion, NH_4^+. This ion consists of a group of five atoms, one nitrogen and four hydrogens, with the whole group of atoms having a single positive charge. Similarly, the carbonate ion, CO_3^{2-}, consists of a group of four atoms, one carbon atom and three oxygen atoms in this case. The whole group of atoms has a double negative charge.

The valency of a group ion is the same as the number of charges it has. For example, the ammonium ion, NH_4^+, with one positive charge, has a valency of one. The carbonate ion, CO_3^{2-}, with two negative charges has a valency of two. Simple chemical formulae for compounds that contain group ions can be worked out using the valency number method.

Example 1
Work out the simple chemical formula for ammonium chloride.

Step 1	symbols	NH_4 Cl
Step 2	valencies	1↘ ↙1

Step 3	cross over valencies	$(NH_4)_1Cl_1$
Step 4	cancel out any common factor	$(NH_4)_1Cl_1$
Step 5	omit '1' if present	NH_4Cl

The simple chemical formula for ammonium chloride is NH_4Cl.

Example 2
Work out the simple chemical formula for magnesium carbonate.

Step 1	symbols	Mg CO_3	
Step 2	valencies	2↘ ↙2	(magnesium is group 2)

Step 3	cross over valencies	$Mg_2(CO_3)_2$
Step 4	cancel out any common factor	$Mg_1(CO_3)_1$
Step 5	omit '1' if present	$MgCO_3$

The simple chemical formula for magnesium carbonate is $MgCO_3$.

Example 3
Work out the simple chemical formula for ammonium carbonate.

Step 1	symbols	NH_4 CO_3
Step 2	valencies	1↘ ↙2

Step 3	cross over valencies	$(NH_4)_2(CO_3)_1$
Step 4	cancel out any common factor (not required in this case)	$(NH_4)_2(CO_3)_1$
Step 5	omit '1' if present	$(NH_4)_2CO_3$

The simple chemical formula for ammonium carbonate is
$$(NH_4)_2CO_3$$

Question

4 In each of the compounds given below there is at least one group ion. You will not need any brackets in your final answer. Write simple chemical formulae for:

(a) lithium hydroxide
(b) potassium carbonate
(c) calcium sulphate
(d) strontium carbonate
(e) potassium chromate
(f) sodium sulphite
(g) lithium hydrogencarbonate
(h) sodium nitrate
(i) potassium phosphate

(j) beryllium chromate
(k) sodium dichromate
(l) ammonium bromide
(m) lithium hydrogensulphite
(n) potassium sulphate
(o) aluminium phosphate
(p) ammonium nitrite
(q) magnesium sulphite
(r) sodium carbonate

Simple formulae for compounds using Roman numerals and brackets

Simple formulae from Roman numerals

Some elements, such as transition metals, can have more than one valency, for example iron can have a valency of 2 or 3. Chemists use Roman numerals in the names of compounds to show which valency is being used. For example, iron forms iron(II) oxide, in which the iron has a valency of two and iron(III) oxide, in which it has a valency of three.

Example 1

Work out the simple chemical formula for iron(II) oxide.

Step 1 symbols	Fe	O
Step 2 valencies	2	2

Step 3 cross over valencies	Fe_2O_2
Step 4 cancel out any common factor	Fe_1O_1
Step 5 omit '1' if present	FeO

The simple chemical formula for iron(II) oxide is FeO.

Example 2

Work out the simple chemical formula for iron(III) oxide.

Step 1 symbols	Fe	O
Step 2 valencies	3	2

Step 3 cross over valencies	Fe_2O_3
Step 4 cancel out any common factor	Fe_2O_3
Step 5 omit '1' if present	Fe_2O_3

The simple chemical formula for iron(III) oxide is Fe_2O_3.

Simple formulae containing brackets

If more than one of the same group ion is present in a chemical formula, then brackets are placed around that ion, followed by a number to indicate how many of those ions are present. For example, the formula $Ca(OH)_2$, indicates that, in calcium hydroxide, there are two hydroxide ions, OH^-, for every calcium ion, Ca^{2+}.

Example 3

Work out the simple chemical formula for calcium hydroxide.

Step 1 symbols	Ca	OH
Step 2 valencies	2	1

Step 3 cross over valencies	$Ca_1(OH)_2$
Step 4 cancel out any common factor	$Ca_1(OH)_2$
Step 5 omit '1' if present	$Ca(OH)_2$

The simple chemical formula for calcium hydroxide is $Ca(OH)_2$.

Some Roman numerals used in chemistry

I = 1
II = 2
III = 3
IV = 4
V = 5
VI = 6
VII = 7

Sometimes, when writing a chemical formula for a compound, we require knowledge of both Roman numerals and the use of brackets.

Example 4

Work out the simple chemical formula for iron(III) hydroxide.

Step 1 symbols   Fe   OH

Step 2 valencies   3   1

Step 3 cross over valencies $Fe_1(OH)_3$

Step 4 cancel out any common factor $Fe_1(OH)_3$

Step 5 omit '1' if present $Fe(OH)_3$

The simple chemical formula for iron(III) hydroxide is $Fe(OH)_3$.

Questions

5 The following compounds contain only two elements. One is a metal with the valency shown by a Roman numeral, the other is a non-metal. Write simple chemical formulae for:

(a) iron(II) chloride   (h) silver(I) iodide
(b) nickel(II) sulphide   (i) bismuth(III) sulphide
(c) manganese(IV) oxide   (j) silver(II) fluoride
(d) tin(IV) chloride   (k) copper(I) oxide
(e) lead(II) iodide   (l) uranium(VI) fluoride
(f) lead(II) oxide   (m) manganese(VII) oxide
(g) antimony(V) oxide   (n) bismuth(III) hydride

6 The following compounds all contain group ions. You will need brackets in your answer. Write simple chemical formulae for:

(a) magnesium hydroxide   (h) ammonium sulphite
(b) barium phosphate   (i) barium hydrogensulphate
(c) beryllium nitrate   (j) aluminium hydroxide
(d) calcium hydrogencarbonate   (k) aluminium sulphate
(e) ammonium phosphate   (l) strontium nitrite
(f) magnesium nitrate   (m) calcium hydrogensulphite
(g) calcium nitrite   (n) ammonium carbonate

7 In these compounds you require knowledge of the use of Roman numerals and/or the use of brackets. Write simple chemical formulae for:

(a) silver(I) nitrate   (h) iron(II) hydroxide
(b) zinc(II) nitrate   (i) silver(I) carbonate
(c) magnesium permanganate   (j) copper(II) sulphate
(d) nickel(II) chloride   (k) mercury(II) nitrate
(e) chromium(III) sulphate   (l) lead(IV) fluoride
(f) ammonium dichromate   (m) vanadium(V) oxide
(g) manganese(II) nitrate   (n) aluminium chromate

Ionic formulae for compounds

Ionic formulae for two-element compounds

Ionic compounds conduct electricity when molten and have high melting points. Solutions of ionic compounds also conduct electricity. Ammonium compounds and compounds which contain a metal should be assumed to be ionic, *unless* there is evidence to the contrary. For example, if the compound has a low melting point or does not conduct electricity when molten.

When forming ionic compounds, metals in groups 1, 2 and 3 lose their outer electrons (see page 45). By doing this they obtain the stable electron arrangement of a noble gas. The non-metals in groups 5, 6 and 7 gain electrons and also obtain a stable noble gas electron arrangement. This means that atoms of these non-metals have the number of outer electrons in their outer shell made up to eight.

In group 4, the non-metals carbon and silicon do not form ions based on single atoms. Also in the same group, the metals tin and lead exhibit variable valency like transition metals.

The charges on the ions of main group elements can be summarised as follows:

Group number	1	2	3	4	5	6	7
Charge on ion	1+	2+	3+	/	3−	2−	1−

Example 1
Work out the ionic formula for calcium fluoride (which contains two main group ions.)

Step 1 symbols	Ca F
Step 2 ions	Ca^{2+} F^-
Step 3 number of ions to balance charges	$1Ca^{2+}2F^-$ (2+ balances 2−)
Step 4 ionic formula	$Ca^{2+}(F^-)_2$

It should be noted that although brackets are not needed in the simple chemical formula for calcium fluoride, CaF_2, they are needed in the ionic formula $Ca^{2+}(F^-)_2$. Brackets are always needed in an ionic formula where more than one ion of a particular type is present.

Ionic formulae from Roman numerals

As we already know, a Roman numeral can be used to show the valency of a metal atom. For example, in copper(II) chloride the valency of copper is two. The Roman numeral also gives the size of the charge on the copper ion that is present, which in the case of any copper(II) compound is Cu^{2+}. Similarly, in any copper(I) compound, the valency of copper is one and the copper ion present is Cu^+.

Example 2

Work out the ionic formula for copper(II) chloride.

Step 1 ions $\qquad$ Cu^{2+} $\quad Cl^-$

Step 2 number of ions to balance $\quad 1Cu^{2+}$ $\quad 2Cl^-$ $\quad$ (2+ balances 2−)
charges

Step 3 ionic formula $\qquad$ $Cu^{2+}(Cl^-)_2$

The ionic formula for copper(II) chloride is $Cu^{2+}(Cl^-)_2$

Example 3

Work out the ionic formula for nickel(II) nitrate. (The formula for a group ion such as the nitrate ion is given in the data booklet.)

Step 1 ions $\qquad$ Ni^{2+} $\quad NO_3^-$

Step 2 number of ions to $\qquad$ $1Ni^{2+}$ $\quad 2NO_3^-$ $\quad$ (2+ balances 2−)
balance charges

Step 3 ionic formula $\qquad$ $Ni^{2+}(NO_3^-)_2$

The ionic formula for nickel(II) nitrate is $Ni^{2+}(NO_3^-)_2$

Example 4

Work out the ionic formula for chromium(III) sulphate. (The formula for the sulphate ion is given in the data booklet.)

Step 1 ions $\qquad$ Cr^{3+} $\quad SO_4^{2-}$

Step 2 number of ions to $\qquad$ $2Cr^{3+}$ $\quad 3SO_4^{2-}$ $\quad$ (6+ balances 6−)
balance charges

Step 3 ionic formula $\qquad$ $(Cr^{3+})_2(SO_4^{2-})_3$

The ionic formula for chromium(III) sulphate is $(Cr^{3+})_2(SO_4^{2-})_3$

Questions

8 The following compounds are made up of two main group elements. Write ionic formulae for:

(a) potassium sulphide
(b) barium fluoride
(c) sodium fluoride
(d) calcium oxide
(e) lithium nitride
(f) strontium phosphide
(g) aluminium oxide
(h) magnesium bromide
(i) calcium sulphide
(j) beryllium chloride
(k) aluminium fluoride
(l) lithium oxide
(m) potassium nitride
(n) sodium sulphide

9 In the following compounds at least one group ion is present, but no brackets are needed. Write ionic formulae for:

(a) lithium hydroxide
(b) potassium nitrate
(c) barium carbonate
(d) ammonium fluoride

(e) lithium hydrogencarbonate (j) ammonium nitrate
(f) sodium hydrogensulphite (k) calcium sulphite
(g) calcium chromate (l) strontium dichromate
(h) magnesium sulphate (m) caesium hydrogensulphate
(i) lithium permanganate (n) beryllium carbonate

10 In the following compounds at least one group ion is present. Brackets are needed in each case. Write ionic formulae for:

(a) potassium carbonate (h) calcium hydrogencarbonate
(b) lithium sulphate (i) sodium sulphite
(c) ammonium sulphate (j) strontium nitrite
(d) sodium chromate (k) magnesium hydrogensulphite
(e) lithium dichromate (l) ammonium phosphate
(f) aluminium hydroxide (m) radium nitrate
(g) barium phosphate (n) aluminium sulphate

11 Write ionic formulae for:

(a) copper(II) fluoride (h) strontium hydrogensulphate
(b) zinc(II) sulphide (i) ammonium hydrogensulphate
(c) nickel(II) hydroxide (j) calcium permanganate
(d) silver(I) carbonate (k) chromium(III) sulphate
(e) chromium(III) oxide (l) magnesium phosphate
(f) copper(II) nitrate (m) manganese(II) nitrate
(g) iron(III) hydroxide (n) aluminium nitrate

Chemical equations

Word equations

In all chemical reactions one or more new substances are formed. For example, in many school laboratories mains gas is burned in a Bunsen burner as a source of heat in experiments (see Figure 5.1). The initial substances in the reaction, called the **reactants**, are methane and oxygen. The substances formed in the reaction, called the **products**, are carbon dioxide and water vapour. The Bunsen flame is where the reaction between the methane and oxygen to produce carbon dioxide and water takes place. A **word equation** can be written for any chemical reaction that shows the names of the reactants and the products. The word equation for the reaction taking place in the Bunsen flame is:

$$\text{methane} + \text{oxygen} \rightarrow \text{carbon dioxide} + \text{water}$$

In all chemical equations, including word equations, '+' means 'and' and '→ ' means 'react(s) to produce'. So the word equation above tells us that methane and oxygen react to produce carbon dioxide and water.

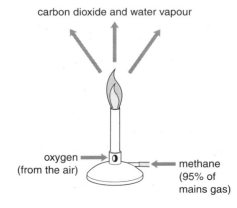

carbon dioxide and water vapour

oxygen (from the air)

methane (95% of mains gas)

Figure 5.1 ▲
A Bunsen burner.

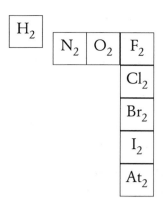

Figure 5.2 ▲

Formula equations

Any word equation can be re-written as a **formula equation** using chemical symbols and formulae for reactants and products. It is normal to use simple chemical formulae for ionic and covalent network compounds. For example, sodium chloride would be shown as NaCl and the covalent network silicon dioxide as SiO_2. For covalent molecular compounds, the molecular formula is used. In the case of methane this is CH_4. Almost all elements are represented by a single symbol, for example C for carbon and Zn for zinc. However, *diatomic* elements are represented by their molecular formula, for example hydrogen is shown as H_2. This means that, for the diatomic elements only, a subscript '2' is written after the symbol. All of the diatomic elements are shown as molecular formulae in Figure 5.2.

Example 1
When hydrogen burns, it combines with oxygen to produce water.

Word equation: hydrogen + oxygen → water

Formula equation: $H_2 + O_2 \rightarrow H_2O$

Example 2
When methane burns completely, it combines with oxygen to produce carbon dioxide and water.

Word equation: methane + oxygen → carbon dioxide + water

Formula equation: $CH_4 + O_2 \rightarrow CO_2 + H_2O$

Example 3
Calcium reacts with water to give calcium hydroxide and hydrogen.

Word equation: calcium + water → calcium hydroxide + hydrogen

Formula equation: $Ca + H_2O \rightarrow Ca(OH)_2 + H_2$

The formula equations shown above are 'unbalanced'. The balancing of formula equations is dealt with in the next section.

Question

12 Write word equations and unbalanced formula equations for:

  (a) sodium reacting with chlorine to give sodium chloride
  (b) hydrogen reacting with bromine to give hydrogen bromide
  (c) magnesium reacting with oxygen to give magnesium oxide
  (d) calcium carbonate decomposing to give calcium oxide and carbon dioxide
  (e) potassium hydroxide reacting with carbon dioxide to give potassium carbonate and water
  (f) lithium oxide reacting with water to give lithium hydroxide

(g) barium nitrate and sodium sulphate reacting to produce
    barium sulphate and sodium nitrate
(h) zinc and copper(II) chloride reacting to produce copper and
    zinc(II) chloride.

Balancing chemical equations

Chemists often have to carry out calculations, for example to find out the mass of a product formed in a reaction. To do this they need a **balanced chemical equation** for each reaction.

A chemical equation is said to be balanced when there are equal numbers of each type of atom on both sides of the equation. This is logical since atoms are not made or destroyed in a chemical reaction. All of the atoms that were present in the reactants are still present in the products.

Chemical equations should only be balanced by putting numbers *in front of* formulae. Formulae for reactants and products should not be altered in any way when balancing equations. It is usually advisable to start with the first element on the left hand side of the equation and continue methodically through it. It may be necessary to go through the equation more than once, so it is important to carry out a final check to make sure that you have not unbalanced any elements.

Example 1
In the complete combustion of methane, methane and oxygen react to produce carbon dioxide and water. The *unbalanced* formula equation for this is:

$$CH_4 + O_2 \rightarrow CO_2 + H_2O$$

Although there is one carbon atom on both sides of the equation, there are four hydrogen atoms on the left, but only two on the right. There are also two oxygen atoms on the left, but three on the right. This can be seen more clearly if full structural formulae are used.

Left:			Right:		
	H atoms	4		H atoms	2
	C atoms	1		C atoms	1
	O atoms	2		O atoms	3

The four hydrogen atoms in the methane molecule will produce two water molecules, each containing two hydrogen atoms. Two

oxygen molecules will be required to provide the four oxygen atoms that are present in the products.

Left: H atoms   4   Right: H atoms   4
        C atoms   1   C atoms   1
        O atoms   4   O atoms   4

The balanced chemical equation can now be written as:

$$CH_4 + 2O_2 \rightarrow CO_2 + 2H_2O$$

Example 2
Balance the equation:

$$H_2 + O_2 \rightarrow H_2O$$

In this equation there are two hydrogen atoms on each side and therefore these are balanced, so proceed to oxygen. There are two oxygen atoms on the left, but only one on the right, so we put a 2 in front of H_2O. This gives:

$$H_2 + O_2 \rightarrow 2H_2O$$

Checking again, we find that there are now two hydrogen atoms on the left, but four on the right. In order to have four hydrogen atoms on the left we put a 2 in front of H_2. This gives:

$$2H_2 + O_2 \rightarrow 2H_2O$$

A final check shows that the equation is now balanced:

Left: H atoms   4   Right: H atoms   4
        O atoms   2   O atoms   2

Example 3
Sometimes the use of a fraction, such as ½, can help to balance an equation. Balance the equation:

$$C_2H_6 + O_2 \rightarrow CO_2 + H_2O$$

Balancing the carbon atoms gives:

$$C_2H_6 + O_2 \rightarrow 2CO_2 + H_2O$$

Balancing the hydrogen atoms gives:

$$C_2H_6 + O_2 \rightarrow 2CO_2 + 3H_2O$$

There are now seven oxygen atoms on the right, so the number put in front of O_2 must give seven when multiplied by the 2 in O_2.

This number is 3½, which gives:

$$C_2H_6 + 3½O_2 \rightarrow 2CO_2 + 3H_2O$$

Left:			Right:		
	C atoms	2		C atoms	2
	H atoms	6		H atoms	6
	O atoms	7		O atoms	7

The equation is now balanced and is quite acceptable. However, multiplying right through by two gives the simplest *whole* numbers possible, which some might prefer. The equation then becomes:

$$2C_2H_6 + 7O_2 \rightarrow 4CO_2 + 6H_2O$$

Left:			Right:		
	C atoms	4		C atoms	4
	H atoms	12		H atoms	12
	O atoms	14		O atoms	14

Example 4

There are rare occasions when the methodical approach described in the earlier examples does not lead to a balanced equation. In such cases a 'trial and error' method may be adopted, or it may be possible to solve the problem by a closer examination of the particular equation.

Consider balancing the equation:

$$Fe_2O_3 + CO \rightarrow Fe + CO_2$$

Balancing the iron atoms gives:

$$Fe_2O_3 + CO \rightarrow 2Fe + CO_2$$

Trying to balance the carbon and oxygen atoms appears to be impossible by the usual method. However, it should be noticed that each carbon monoxide molecule removes one oxygen atom from the iron(III) oxide in forming carbon dioxide. There are three oxygen atoms to be removed and this therefore requires three carbon monoxide molecules. In turn, this leads to the production of three carbon dioxide molecules. The balanced equation is therefore:

$$Fe_2O_3 + 3CO \rightarrow 2Fe + 3CO_2$$

Left:			Right:		
	Fe atoms	2		Fe atoms	2
	O atoms	6		O atoms	6
	C atoms	3		C atoms	3

13 Balance the following chemical equations. In each case give the simplest possible *whole numbers* in your final answer. Some equations are already balanced.

(a) $Mg + O_2 \rightarrow MgO$

(b) $K + I_2 \rightarrow KI$

(c) $H_2 + Cl_2 \rightarrow HCl$

(d) $PbO_2 + H_2 \rightarrow Pb + H_2O$

(e) $Mg + H_2SO_4 \rightarrow MgSO_4 + H_2$

(f) $Li + H_2O \rightarrow LiOH + H_2$

(g) $KOH + CO_2 \rightarrow K_2CO_3 + H_2O$

(h) $Cl_2 + NaBr \rightarrow NaCl + Br_2$

(i) $AgNO_3 \rightarrow Ag + NO_2 + O_2$

(j) $C_2H_4 + O_2 \rightarrow CO_2 + H_2O$

(k) $C_3H_6 + O_2 \rightarrow CO_2 + H_2O$

(l) $K_2CO_3 + HCl \rightarrow KCl + H_2O + CO_2$

(m) $Zn + HNO_3 \rightarrow Zn(NO_3)_2 + H_2$

(n) $CuCO_3 \rightarrow CuO + CO_2$

(o) $FeS_2 + O_2 \rightarrow Fe_2O_3 + SO_2$

(p) $NH_3 + O_2 \rightarrow N_2 + H_2O$

6 The Mole

Formula mass and the mole

Have you ever wondered how the quantities of the chemicals you use in your experiments are worked out? Why, for example, was it important to weigh out 2 g of calcium carbonate for use in a particular experiment and not some other mass? In fact, most quantities are calculated using simple arithmetic. Two important tools for these calculations, formula masses and moles, are looked at in this section.

Formula mass

The **formula mass** of a substance (more correctly called the relative formula mass) is obtained by adding the relative atomic masses of the elements in the formula.

Example 1
Calculate the formula mass of calcium chloride.

Formula: $CaCl_2$
Formula mass: $40 + (35.5 \times 2) = 111$

Notice that the relative atomic mass of chlorine is multiplied by 2.

Example 2
Calculate the formula mass of ammonium sulphate.

Formula: $(NH_4)_2SO_4$
Formula mass: $([14 + (1 \times 4)] \times 2) + 32 + (16 \times 4)$
                  $= (18 \times 2) + 32 + 64 = 132$

Notice that the 2 outside the bracket in $(NH_4)_2$ multiplies the contents of the brackets by 2.

The mole

The (relative) formula masses of substances have no units, but in order to weigh out chemicals we need to know the units. Chemists use a special term called the **mole** which effectively gives the units of grams to formula masses.

A **mole** of a substance is the formula mass expressed in grams. For example, the mass of one mole of calcium chloride is 111 g and the mass of one mole of ammonium sulphate is 132 g.

A mole of a substance is often referred to as the 'gram formula mass' and given the abbreviation GFM.

For any element or compound, the mass present can be calculated from knowledge of the number of moles and the formula mass.

Mass of substance = number of moles × mass of one mole
= number of moles × formula mass in grams

Mass of substance in grams = number of moles × formula mass

Example 3
Calculate the mass of 2.5 moles of calcium carbonate, $CaCO_3$.

$$\text{Mass of } CaCO_3 \text{ in grams} = \text{number of moles} \times \text{formula mass}$$
$$= 2.5 \times [40 + 12 + (16 \times 3)] = 250 \, g$$

Rearranging the relationship used above allows the number of moles of a substance to be calculated, if the mass in grams and the formula mass are known.

$$\text{Number of moles of substance} = \frac{\text{mass of substance in grams}}{\text{formula mass}}$$

Example 4
Calculate the number of moles of water in 100 g of water.

$$\text{Number of moles of } H_2O = \frac{\text{mass of water in grams}}{\text{formula mass}}$$
$$= \frac{100}{[(1 \times 2) + 16]} = \frac{100}{18} = 5.56$$

The same relationship can also be rearranged to allow the calculation of the formula mass of a substance if the mass in grams and the number of moles are known.

$$\text{Formula mass of substance} = \frac{\text{mass of substance in grams}}{\text{number of moles of substance}}$$

Example 5
Three moles of gas X have a mass 132 g. Calculate the formula mass of gas X and suggest what it might be.

$$\text{Formula mass of X} = \frac{\text{mass of X in grams}}{\text{number of moles of X}}$$

$$= \frac{132}{3} = 44 \text{ (possibly carbon dioxide, } CO_2)$$

Summary

The relationship between the *mass* of any pure substance in grams (m), the number of *moles* of that substance (n) and its *formula mass* (FM), can be summed up in a triangle of knowledge.

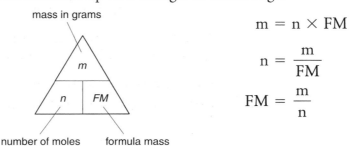

mass in grams

number of moles   formula mass

$$m = n \times FM$$

$$n = \frac{m}{FM}$$

$$FM = \frac{m}{n}$$

1 Calculate the (relative) formula mass of each of the following:

(a) nitrogen, N_2
(b) methane, CH_4
(c) potassium hydroxide, KOH
(d) magnesium hydroxide, $Mg(OH)_2$
(e) copper(II) sulphate, $CuSO_4$
(f) ammonium carbonate, $(NH_4)_2CO_3$
(g) lithium phosphate, Li_3PO_4

2 Calculate the mass of each of the following:

(a) 1 mole of argon, Ar
(b) 1 mole of magnesium nitride, Mg_3N_2
(c) 1 mole of potassium nitrate, KNO_3
(d) 6 moles of propane, C_3H_8
(e) 3.5 moles of sodium sulphate, Na_2SO_4
(f) 0.25 moles of ammonium nitrate, NH_4NO_3
(g) 0.016 moles of calcium iodate, $Ca(IO_3)_2$

3 Calculate the number of moles in each of the following:

(a) 15 g of hydrogen, H_2
(b) 90 g of ethane, C_2H_6
(c) 850 g of silver(I) nitrate, $AgNO_3$
(d) 250 g of ammonium chloride, NH_4Cl
(e) 2.25 kg of ammonium sulphate, $(NH_4)_2SO_4$ (1 kg = 1000 g)
(f) 24 kg of calcium nitrate, $Ca(NO_3)_2$
(g) 1.80 g of glucose, $C_6H_{12}O_6$

4 Calculate the formula mass of each of the following:

(a) 2 moles of a substance that has a mass of 36 g
(b) 6 moles of a substance that has a mass of 180 g
(c) 4.4 moles of a substance that has a mass of 466.4 g
(d) 12.5 moles of a substance that has a mass of 925 g
(e) 8.7 moles of a substance that has a mass of 2.697 kg

5 Give possible names and formulae for the substances referred to in 4(a) and 4(b).

Calculations based on balanced equations

Balanced equations can be used to make predictions about the masses of substances either reacting or being produced. In order to do this, every formula is taken to represent one mole of the substance concerned. The abbreviation that chemists use for the word 'mole' is 'mol'.

A calculation based on a balanced equation can be broken down into five stages:

1 Write a balanced equation for the reaction.

2 Identify the number of moles of the substances concerned.

3 Replace moles by formula masses in grams.

4 Change grams to any other unit of mass if necessary.

5 Use simple proportion to complete the calculation.

Example 1 (no change of units required)
Calculate the mass of magnesium oxide produced when 6 g of magnesium burns in air or oxygen.

$$2Mg + O_2 \longrightarrow 2MgO$$

2 mol $\longleftrightarrow$ 2 mol

49 g $\longleftrightarrow$ 81 g

6 g $\longleftrightarrow$ $\dfrac{81 \times 6}{49} = 9.92\,g$

Example 2 (change of units required)
What mass of water is produced when 10 tonnes of methane burns completely in air?

$$CH_4 + 2O_2 \rightarrow CO_2 + 2H_2O$$

1 mol $\longleftrightarrow$ 2 mol

16 g $\longleftrightarrow$ 36 g

16 tonnes $\longleftrightarrow$ 36 tonnes

10 tonnes $\longleftrightarrow$ $\dfrac{36 \times 10}{16} = 22.5\,tonnes$

Questions

6 Calcium carbonate is used in indigestion remedies to neutralise excess stomach acid (hydrochloric acid). What mass of stomach acid could be neutralised by 1 g of calcium carbonate?

$$CaCO_3 + 2HCl \rightarrow CaCl_2 + H_2O + CO_2$$

7 (a) Lithium hydroxide is used to absorb carbon dioxide in spacecraft. What mass of carbon dioxide could be absorbed by 1 kg of lithium hydroxide?

$$2LiOH + CO_2 \rightarrow Li_2CO_3 + H_2O$$

(b) What mass of carbon dioxide could be absorbed by 1 kg of sodium hydroxide?

$$2NaOH + CO_2 \rightarrow Na_2CO_3 + H_2O$$

(c) Why is lithium hydroxide used to absorb carbon dioxide in spacecraft rather than sodium hydroxide, which costs less?

8 The fertiliser ammonium phosphate is made by reacting ammonia with phosphoric acid. What mass of ammonium phosphate can be produced from 250 kg of ammonia?

$$3NH_3 + H_3PO_4 \rightarrow (NH_4)_3PO_4$$

9 What mass of hydrogen is produced when 2 g of magnesium reacts with an excess of dilute sulphuric acid?

$$Mg + H_2SO_4 \rightarrow MgSO_4 + H_2$$

10 What mass of ammonia, NH_3, is produced when 5 g of ammonium sulphate reacts with an excess of sodium hydroxide?

$$(NH_4)_2SO_4 + 2NaOH \rightarrow Na_2SO_4 + 2H_2O + 2NH_3$$

11 A coal-burning power station produces 70 tonnes of sulphur dioxide in one day. What mass of sulphuric acid, H_2SO_4, is produced in the atmosphere if it reacts with oxygen and water as follows?

$$2SO_2 + O_2 + 2H_2O \rightarrow 2H_2SO_4$$

12 In one year a family burns 2100 kg of propane, C_3H_8, in their central heating boiler. What mass of carbon dioxide is produced?

$$C_3H_8 + 5O_2 \rightarrow 3CO_2 + 4H_2O$$

Carbon Compounds

7 Fuels

Combustion

Figure 7.1 ▲
Fuels provide energy for heat, light and transport.

What do coal, wood, diesel, petrol and natural gas have in common? They are all substances that can be burned to produce energy and, because of this, are called **fuels**. Fuels provide energy for heat, light and transport as shown in Figure 7.1.

Fuels supply heat for homes and for cooking. Coal and natural gas are used in many power stations to make electricity. Diesel and petrol are used in cars, lorries and buses.

When they burn and give out energy fuels combine with oxygen in a chemical reaction which chemists call **combustion**. Combustion is therefore another word for burning.

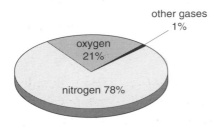

Figure 7.2 ▲
Composition of air.

Air contains oxygen and other gases. The pie chart in Figure 7.2 shows that air contains 78% nitrogen, 21% oxygen and 1% of other gases. As a rough guide, we can say that air is made up of one-fifth oxygen, the rest being mainly nitrogen.

If the air contained more oxygen, there would be an increased risk of forest fires. This is because pure oxygen, or air enriched with oxygen, can cause a glowing wood splint to re-light. When a glowing splint is placed in a test tube containing oxygen it re-lights and burns brightly (see Figure 7.3). Since oxygen is the only common gas that can do this, it is used as the chemical test for this gas. We usually say that 'oxygen re-lights a glowing splint'.

As was said earlier in Chapter 1, all combustion reactions are **exothermic**. That is to say, they give out heat energy. The reason why fuels are important is that, when they burn, the reactions are very exothermic and give out a lot of heat energy. This, in turn, may be turned into other forms of energy such as electrical energy or kinetic (movement) energy.

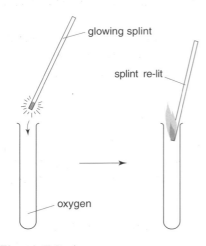

Figure 7.3 ▲
Oxygen re-lights a glowing splint.

Questions

1 What is meant by the term 'fuel'?

2 Which gas is used up when a substance burns?

3 Give another word for burning.

The two main energy sources

Crude oil and **natural gas** currently provide about two-thirds of the world's energy.

Crude oil and natural gas are often found together and are thought to have been formed from the decay of animals and plants which lived in shallow seas many millions of years ago. When they died, they sank to the sea bed and were covered with layers of sediment such as sand and mud. High pressure and temperature resulted in decomposition into the much sought-after crude oil and natural gas that our society finds so useful. The oil and gas are found trapped in porous rock, such as sandstone and limestone. Crude oil is more accurately referred to as **petroleum**, which is derived from the Greek words for rock (petra) and oil (oleum).

Both crude oil and natural gas are mixtures consisting mainly of compounds called **hydrocarbons** that contain only the elements hydrogen and carbon. All hydrocarbons undergo complete combustion, combining with oxygen to produce carbon dioxide and water *if* there is a plentiful supply of air. Methane, the simplest hydrocarbon, has molecular formula CH_4 and is the main constituent of natural gas. Equations for its complete combustion are as follows:

Word equation: methane + oxygen → carbon dioxide + water

Balanced formula equation: $CH_4 + 2O_2 \rightarrow CO_2 + 2H_2O$

The apparatus in Figure 7.4 can be used to show that carbon dioxide and water are produced when a hydrocarbon fuel burns. A liquid hydrocarbon is shown burning, but a gaseous hydrocarbon like methane could also be used if it were burned at the end of a glass or metal tube.

The gases from the burning hydrocarbon are drawn through the apparatus using a pump. Carbon dioxide in the gas mixture soon turns the lime water milky, which is the chemical test for that gas. Condensation is also soon seen in the cooled test tube as water vapour from the burning hydrocarbon condenses. If enough is collected, which takes much longer, the colourless liquid that collects may be proved to be water by showing that it freezes at 0°C and boils at 100°C.

Air contains both water vapour and carbon dioxide and so, as a control experiment, air alone should be drawn through the apparatus for the same length of time to show that the positive results are caused by products of combustion and not by air.

The experiment in Figure 7.4 tells us two things about the atoms present in the hydrocarbon fuel. Because carbon dioxide is produced, the fuel must contain carbon atoms. Also, because water (hydrogen oxide) is produced, the fuel must contain hydrogen atoms. Only the oxygen atoms in the compounds carbon dioxide and water came from the air.

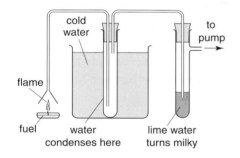

Figure 7.4 ▲
Apparatus to show that combustion of a hydrocarbon gives carbon dioxide and water.

Questions

4 Which two-element compounds make up most of crude oil and natural gas?

5 Write a balanced formula equation for the complete combustion of propane.

6 Describe the tests for (a) carbon dioxide and (b) water.

The greenhouse effect

One of the reasons why the Earth's atmosphere is so important is that it acts as a huge blanket, keeping the Earth reasonably warm, even at night. Some of the gases in the atmosphere are better at keeping the Earth warm than others. Those that do are called '**greenhouse gases**' because, just like the glass in a greenhouse, they allow through the light from the sun, but they absorb the longer wavelength radiation emitted by the warmed surface of the Earth. Some of this energy they then re-radiate back to the Earth (see Figure 7.5).

The most important of the greenhouse gases is carbon dioxide. It is believed that for about 10 000 years the level of carbon dioxide in the air remained roughly constant at about 0.027%. However, with increasing use of carbon-containing fuels, such as coal, petrol, diesel and natural gas, the amount of carbon dioxide has risen to about 0.035% and continues to rise. There is now clear evidence of global warming, caused by the increase in carbon dioxide levels, since average surface temperatures all over the Earth are rising. We see this in the melting of polar ice caps and glaciers, which will result in a rise in sea levels that will eventually submerge low-lying land.

Carbon dioxide is not the only greenhouse gas. Other gases that also cause global warming include water vapour and methane. The levels of both of these gases in the atmosphere are also increasing.

Questions

7 Name the most important 'greenhouse gas'.

8 How are increasing levels of greenhouse gases in the atmosphere likely to affect the surface temperature of the Earth?

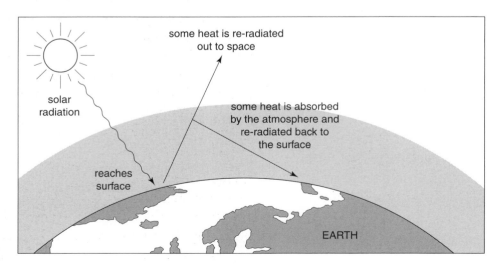

Figure 7.5 ▲
The greenhouse effect.

Fuels and pollution

A **pollutant** is a substance which harms the environment and therefore carbon dioxide and the other gases that cause global warming can be considered to be pollutants. The combustion of fuels also produces many other pollutants.

Carbon monoxide

Any fuel that contains carbon can produce carbon monoxide when it burns if the combustion takes place with insufficient oxygen present to give carbon dioxide. Unlike carbon dioxide, carbon monoxide is a very poisonous gas since it combines with haemoglobin in the blood, which is the body's oxygen carrier. Starved of oxygen, a person breathing in the colourless and odourless carbon monoxide can soon be rendered unconscious and may die. All petrol and diesel engines produce carbon monoxide.

Carbon particles

The carbon particles, known as 'soot', that are produced by the combustion of the hydrocarbons in diesel are known to be harmful. Efficient filters should prevent them escaping from exhaust pipes but studies of taxi drivers working in cities have shown increased ill health which has been attributed to soot emissions from diesel engines.

Sulphur dioxide

Coal, crude oil and natural gas all contain varying amounts of sulphur compounds which, when they burn, produce the gas sulphur dioxide. This is a poisonous and irritating gas which is particularly harmful to asthma sufferers. It dissolves in rain water, eventually producing sulphuric acid. The rain water is then called **acid rain**, the harmful effects of which are dealt with in Section 3.

The oil companies make great efforts to remove as much of the sulphur as possible from petrol and diesel during the refining processes, although some sulphur is still present in the final products. Likewise, natural gas has most of the sulphur compounds removed before sale to customers. The same is not true of coal, however, although at some coal-fired power stations, such as Drax in Yorkshire, sulphur dioxide is removed from the flu gases in a process called flu-gas desulphurisation. Unfortunately, for every molecule of poisonous sulphur dioxide removed, a molecule of the greenhouse gas carbon dioxide is produced.

Oxides of nitrogen

In petrol engines, and to a lesser extent in diesel engines, nitrogen and oxygen from the air react to form poisonous oxides of nitrogen, including the reddish-brown acidic gas nitrogen dioxide. Nitrogen and oxygen do not normally react in the air because nitrogen is very unreactive, but they can react if enough energy is supplied in the form of a high temperature.

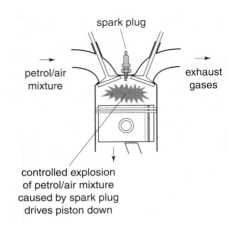

Figure 7.6 ▲
A petrol engine.

In a petrol engine spark plugs provide electrical sparks to ignite a mixture of petrol vapour and air. In the region of the spark the temperature is so high that not only do petrol and oxygen molecules react, but nitrogen and oxygen molecules react as well. The exhaust gases that leave the engine thus contain oxides of nitrogen. How a petrol engine works is shown in Figure 7.6.

Diesel engines do not have spark plugs and do not reach such high temperatures as petrol engines. They therefore cause less reaction between nitrogen and oxygen molecules.

Like sulphur dioxide, nitrogen dioxide dissolves in rain water and contributes to acid rain, as it reacts with water and oxygen to give nitric acid. Rain water is naturally slightly acidic due to dissolved carbon dioxide, but acid rain is unusually acidic due to the presence of sulphuric acid and nitric acid. There is more sulphuric acid than nitric acid present in acid rain, therefore sulphur dioxide contributes more to this kind of pollution than nitrogen dioxide.

Catalytic converters – helping to reduce pollution

All new cars are fitted with **catalytic converters** in their exhaust systems and these, when hot, help to reduce some of the harmful emissions from car engines (see page 24 in Section 1). Helped by the presence of transition metal catalysts, such as platinum, they convert pollutant gases into less harmful ones – oxides of nitrogen into nitrogen, carbon monoxide into carbon dioxide and unburned hydrocarbons into water vapour. However, it should be remembered that both carbon dioxide and water vapour are greenhouse gases.

The hydrocarbon mixture that makes up petrol is quite complex, with different types and sizes of molecules present. One particular hydrocarbon which, in 1998, made up 2.4% of normal unleaded petrol is benzene, a known carcinogen (cancer-causing substance). Because of its harmful nature, it is undesirable for it to be present either in petrol or as an unburned hydrocarbon emitted in the exhaust gases. At refineries efforts are being made to remove as much benzene as possible from petrol because, in 2000, the limit for benzene in all kinds of petrol became 1%.

The need for alternative energy sources

Although crude oil and natural gas are currently being extracted from beneath the surface of the Earth in large quantities, both are **finite** sources of energy and are expected to run out within a hundred years. New sources of energy will have to be found to replace them. Some, such as hydrogen, can be used in motor vehicles like cars and buses. Others, like wind and tidal power, can be used to generate electricity.

9 Name a poisonous gas that is produced by the incomplete combustion of carbon-containing fuels.

10 Name two gases that are responsible for acid rain.

11 Why do petrol engines produce more oxides of nitrogen than diesel engines?

12 Petrol engines produce the pollutants carbon monoxide, oxides of nitrogen and unburned hydrocarbons. Which three less harmful gases are these turned into using a catalytic converter?

Figure 7.7 ▲
This windfarm near Edinburgh has 26 turbines like these which can produce enough electricity for 14 000 homes without pollution.

It is estimated that supplies of coal may last for more than two hundred years and the technology already exists to convert it into liquid fuels that could be used in place of petrol and diesel for example. However, as with the use of hydrogen, this would lead to the production of yet more greenhouse gases, so it may not be desirable use up all of the coal in this way.

One of the best ways of reducing pollution is simply to use *less* energy.

Fractional distillation of crude oil

Crude oil is a complex mixture, made up mainly of hydrocarbons, that is almost useless until it has been refined. However, once refined, it provides almost half of the world's current energy needs and most of its organic (carbon-based) chemicals. Crude oil is therefore very important both as a source of *fuel* and as a source of *chemical feedstocks* for the petrochemicals industry.

The first stage in the refining of crude oil is **fractional distillation**, a continuous process based on differences in boiling points during which it is separated into **fractions**. A fraction being made up of a group of hydrocarbon molecules that have boiling points within a certain range.

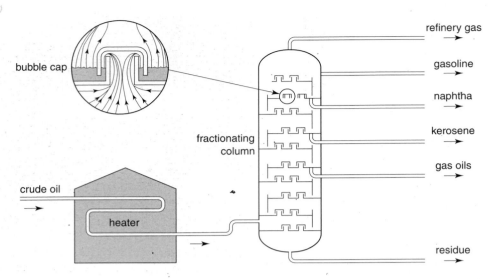

Figure 7.8 ▲
Fractional distillation of crude oil.

Figure 7.9 ▲
Butane gas provides the heat for this barbecue.

Fractional distillation takes place in steel towers about 60 metres high that contain trays with bubble caps (see Figure 7.8). The crude oil is heated to about 350°C before it enters the bottom of the fractionating tower. Some of the crude oil vaporises and moves up the tower, condensing to a liquid where the temperature falls below its boiling point. There is a decrease in temperature as one moves up the tower, but some hydrocarbons remain as gases and leave as the fraction with the lowest boiling point range, **refinery gas**, from the top of the tower. Lower down, four liquid fractions are piped away; these are **gasoline**, **naphtha**, **kerosene** and **gas oil**. The sixth fraction, which does not vaporise, is called the **residue**.

Using the fractions

Refinery gas consists mainly of propane (C_3H_8) and butane (C_4H_{10}). These are either sold as a mixed fuel, or separately, propane usually in red cylinders and butane usually in blue cylinders.

The **gasoline** fraction, consisting mainly of pentane (C_5H_{12}) and hexane (C_6H_{14}) is used to make petrol. It is blended with molecules from the **naphtha** fraction, which have been changed by a process called reforming. The final petrol product consists of hydrocarbons with between 5 and 10 carbon atoms per molecule.

The **naphtha** fraction is not only used in the production of petrol, it is also used for manufacture of a whole range of petrochemicals, including plastics and synthetic fibres. The hydrocarbons in the naphtha fraction contain between 6 and 11 carbon atoms per molecule.

Figure 7.10 ▲
Gasoline and naphtha provide the petrol for this car.

As with the naphtha fraction and the gas oil fraction, sulphur is removed from the **kerosene** fraction after it leaves the fractionating tower. It is used as a fuel in jet aircraft and in domestic heating systems. Kerosene is also called paraffin. Between 9 and 15 carbon atoms are present in the hydrocarbon molecules that make up the kerosene fraction.

Figure 7.11 ▲
Jet aircraft use kerosene as fuel.

Figure 7.12 ▲
Lycra is a synthetic fibre made from naphtha.

Figure 7.14 ▲
Ships like this use heavy fuel oil from the residue fraction.

Figure 7.13 ▲
This train uses diesel fuel from gas oil.

Figure 7.15 ▲
A fractional distillation tower at BP's refinery, Grangemouth.

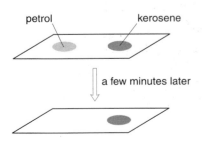

Figure 7.16 ▲
Petrol evaporates more easily than kerosene.

The **gas oil** fraction is used by lorries, buses and trains in the form of diesel engine fuel. Even bigger hydrocarbon molecules are present in this fraction, with each molecule having between 15 and 25 carbon atoms.

Vacuum distillation of the **residue**, which has more than 25 carbon atoms in each of its molecules, gives heavy fuel oil, lubricating oil, wax and bitumen. Ships burn heavy fuel oil in their engines, wax is used to make candles and bitumen is used as tar in making road surfaces.

Summary

Fraction	Boiling range °C	Carbon atoms per molecule	End uses
gas	< 20	1–4	fuel gases
gasoline	20–65	5–6	petrol and petrochemicals
naphtha	65–180	6–11	
kerosene	180–250	9–15	jet fuel/heating
gas oil	250–350	15–25	diesel
residue	> 350	> 25	heavy fuel oil/wax, etc

Properties of the fractions

Boiling range and **viscosity** (how 'thick' the fraction is) increase with increasing molecular size. This is because the forces of attraction between the molecules (the van der Waals' forces – see page 43 of Section 1) increase with increasing molecular size. More energy, and therefore a higher temperature, is needed to separate larger molecules in the liquid state and allow them to move far apart in the gaseous state. Also, as the forces between molecules increase, so the liquid fraction will become less 'runny' and therefore more viscous.

Ease of vaporisation and **flammability** are also linked to molecular size. Smaller hydrocarbon molecules have weaker forces of attraction between them and, as a result, they vaporise more readily than larger molecules. A **flame** is a region where gases react producing heat and

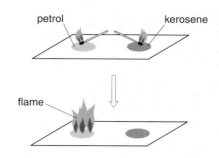

Figure 7.17 ▲
Petrol is more flammable than kerosene.

light. All of the gaseous hydrocarbons are easily ignited with a flame or a spark and the same is true of hydrocarbon liquids that vaporise easily. Figure 7.16 illustrates the point that the smaller hydrocarbon molecules in petrol vaporise more easily than the larger ones in kerosene (paraffin).

A lighted taper easily ignites the vapour above petrol, but not above kerosene (see Figure 7.17). Flammability of the hydrocarbon fractions from crude oil *decreases* as molecular size *increases* and the same is true of ease of vaporisation.

Summary

Fraction	Gas	Gasoline	Naphtha	Kerosene	Gas oil	Residue
molecular size				increasing		→
boiling range				increasing		→
viscosity				increasing		→
flammability				decreasing		→
vaporisability				decreasing		→

The uses of the fractions are related to their ease of vaporisation, viscosity, flammability and boiling range. For example, petrol, kerosene and diesel have low enough viscosity to flow through pipes into the combustion chambers of engines. They are also flammable enough to be ignited in the combustion chambers. Lubricating oil and bitumen from the residue fraction usefully have much higher viscosity and lower flammability, so that lubricating oil 'clings' to moving metal parts and bitumen does not become runny on road surfaces (except during *very* hot weather).

Questions

13 What name is given to the first stage in the refining of crude oil that is based on differences in boiling points?

14 Why do the boiling points of the hydrocarbons present in crude oil increase with increasing molecular size?

15 Naphtha, gas oil and kerosene are fractions from crude oil.

(a) Which fraction has the highest boiling point range?
(b) Which fraction is used for the production of petrol?
(c) Which fraction is used as fuel for jet aircraft?

Chapter 7 Study Questions

In questions 1–4 choose the correct word(s) **from the following list** to complete the sentence.

increases, endothermic, exothermic, petrol, less, decreases, more, diesel

1   The combustion of a fuel is an example of an _____ reaction.

2   With increasing molecular size, the fractions from crude oil become _____ viscous.

3   As the boiling point range of a fraction from crude oil decreases, the flammability _____.

4   Spark plugs are **not** present in a _____ engine.

Questions 5–7 refer to the following end-products from the refining of crude oil:

A   Diesel
B   Bitumen
C   Petrol
D   Kerosene

Which of the above products would be most likely to contain the following hydrocarbon molecules?

5   C_7H_{16}

6   $C_{20}H_{42}$

7   $C_{12}H_{26}$

Questions 8–10 refer to the following gases:

A   Oxygen
B   Carbon dioxide
C   Argon
D   Nitrogen

Which of the above is

8   the most plentiful gas in the atmosphere?

9   a greenhouse gas?

10   the gas used up when fuels burn?

11   Sulphur dioxide is pollutant gas which is a product of the combustion of certain fuels.

   (a) State the meaning of the word 'pollutant'.
   (b) State two properties of sulphur dioxide that cause it to be referred to as a 'pollutant gas'.

12   In the first stage of the refining of crude oil it is separated into fractions.

   (a) What name is given to this method of separation?
   (b) What is meant by the word 'fraction'?

13   One recent estimate of how long the fuels coal, oil and natural gas are likely to last is given below.

Resource	Lifetime/years
Coal	220
Oil	60
Gas	40

Present this information in the form of a bar graph.

14   The table below shows measurements of the carbon dioxide levels in the atmosphere at 5-year intervals.

Year	Carbon dioxide concentration/ parts per million (by volume)
1960	316
1965	319
1970	324
1975	330
1980	337
1985	344
1990	352
1995	359

Present this information as a line graph.

15   Diesel is more viscous than petrol, but is less flammable.

   (a) What is meant by the word 'viscous'?
   (b) Why is diesel more viscous than petrol?
   (c) Why is petrol more flammable than diesel?

16   (a) Why are catalytic converters fitted within the exhaust systems of cars that use petrol as fuel?
   (b) One reaction that occurs within a catalytic converter takes place between nitrogen dioxide and carbon monoxide. The products are two less harmful gases. Write a balanced chemical equation for the reaction.

Hydrocarbons

Alkanes

The **alkanes** (a subset of the set of hydrocarbons) are the main constituents of crude oil and natural gas. They are all covalently bonded molecules, made of carbon and hydrogen atoms, with only single covalent bonds present inside the molecules. In the previous chapter we learned that they are very important both as fuels and as chemical feedstocks for the petrochemicals industry.

In Section 1 we looked in some detail at methane, the simplest of the alkanes, which has molecular formula CH_4. Carbon is, however, unusual among the elements in that it can form long, stable chains of carbon atoms and, as we shall see later, it can also form stable rings. As the alkane molecules increase in size, the forces of attraction between the molecules also increase and, as a result, their boiling points increase.

The eight simplest alkanes are shown in Table 8.1. It should be noticed that alkanes can be recognised by their name ending, which is always **-ane**.

The general formula for the alkanes

If you look at the full structural formulae in Table 8.1 you can see that there is a pattern. Each carbon atom is joined to two hydrogen atoms, except at the ends of the molecules where there are two extra hydrogen atoms. This means that, in every alkane molecule, there are twice as many hydrogen atoms as carbon atoms *and* two more. In mathematical terms, if n is the number of carbon atoms in an alkane molecule, then the number of hydrogen atoms is *2n + 2*. This gives the general molecular formula for any alkane, which is usually called the **general formula** as:

$$C_nH_{2n+2}$$

The general formula can be used to work out the molecular formula of any alkane for which either the number of carbon atoms, or the number of hydrogen atoms is known. For example if the number of carbon atoms, *n*, in an alkane is known to be 10, then the number of hydrogen atoms, *2n + 2*, will be *(2 × 10) + 2 = 22*. The molecular formula for this alkane is therefore $C_{10}H_{22}$ and is, in fact, decane.

You can use the following phrase for remembering the names of the first five alkanes in their correct order:

Must **E**very **Pr**efect **B**e **Pe**rfect

This gives the first letter or letters for methane, ethane, propane, butane and pentane. From pentane onwards the names are based on Green prefixes such as *hexa* for six, *hepta* for seven and *octa* for eight.

Questions

1 What type of bonds are present inside alkane molecules?

2 Write molecular formulae for the following alkanes: **(a)** ethane, **(b)** hexane, **(c)** octane.

3 Write names for the alkanes with the following molecular formulae: **(a)** C_4H_{10}, **(b)** C_5H_{12}, **(c)** C_7H_{16}.

4 Use the general formula for the alkanes to write molecular formulae for: **(a)** nonane, the alkane with 9 carbon atoms, **(b)** dodecane, the alkane with 12 carbon atoms, **(c)** eicosane, the alkane with 42 hydrogen atoms.

5 Give shortened structural formulae for: **(a)** hexane, **(b)** heptane.

6 Write full structural formulae for: **(a)** butane, **(b)** octane.

Name	Molecular formula and state	Full structural formula	Boiling point °C
methane	$CH_4(g)$		−164
ethane	$C_2H_6(g)$		−89
propane	$C_3H_8(g)$		−42
butane	$C_4H_{10}(g)$		−1
pentane	$C_5H_{12}(l)$		36
hexane	$C_6H_{14}(l)$		69
heptane	$C_7H_{16}(l)$		98
octane	$C_8H_{18}(l)$		126

Table 8.1 ▲
The first eight alkanes.

Shortened structural formulae

For the alkanes after methane it is sometimes more convenient to use **shortened structural formulae**. These simply show how many CH_3 and CH_2 groups are present, but still give an indication of the molecular structure. Shortened structural formulae for some of the alkanes are shown in Table 8.2.

Name	Molecular formula	Shortened structural formula
ethane	C_2H_6	CH_3CH_3
propane	C_3H_8	$CH_3CH_2CH_3$
butane	C_4H_{10}	$CH_3CH_2CH_2CH_3$
pentane	C_5H_{12}	$CH_3CH_2CH_2CH_2CH_3$

Table 8.2 ▲
Shortened structural formulae for some alkanes.

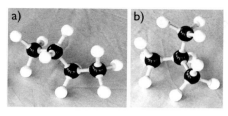

Figure 8.1 ▲

Molecular models of a) butane and b) 2-methylpropane.

Branched chain alkanes

The alkanes like propane and butane are described as **straight chain alkanes** because each carbon atom within their molecules is joined to no more than *two* other carbon atoms. However, crude oil and natural gas both contain another type of alkane called **branched chain alkanes**. In these alkanes some carbon atoms are joined directly to *three* or *four* other carbon atoms. Natural gas that is found above crude oil, for example, contains two different alkanes which both have the molecular formula C_4H_{10}. One is a *straight* chain alkane called butane and the other is a *branched* chain alkane called 2-methylpropane. Their full structural formulae are as follows:

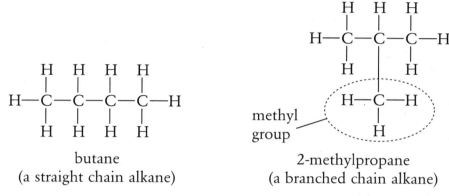

butane
(a straight chain alkane)

2-methylpropane
(a branched chain alkane)

The name 2-methylpropane is given to the branched chain alkane with molecular formula C_4H_{10} because a group of atoms called a methyl group ($-CH_3$) is joined to the *second* carbon atom of a propane molecule in place of one of its hydrogen atoms. Also, one of the carbon atoms in the 2-methylpropane molecule is joined to *three* other carbon atoms.

Rules for naming branched chain alkanes

The rules for naming branched chain alkanes are as follows:

1 Look for the *longest* carbon atom chain to give the name of the parent alkane.

2 Number the carbon atoms in the longest chain from the end that gives the *lowest* numbers to the carbon atoms that have branches attached.

3 Identify the alkyl groups making up the branches. These are likely to be one or more of the following: $-CH_3$ the methyl group, $-CH_2CH_3$ or $-C_2H_5$ the ethyl group, or $-CH_2CH_2CH_3$ the propyl group.

4 The prefixes **di-**, **tri-**, etc are used to indicate how many of a particular side group are present (**di-** = 2, **tri-** = 3, etc).

5 Indicate the position of each side group with a separate number placed in front of its name.

6 If more than one type of side group is present, the names are in alphabetical order. Thus ethyl, for example, is always placed before methyl.

Examples of naming branched chain alkanes

a)

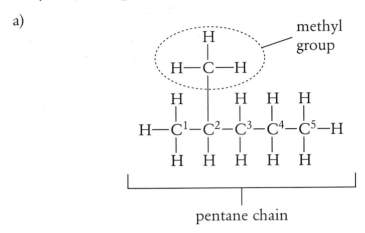

pentane chain

A methyl group is attached to the *second* carbon atom of a pentane chain. The name for this branched chain alkane is **2-methylpentane**. It should be noted that the pentane chain is numbered from the end which produces the *smallest* number, so this compound is *not* named from the right-hand side giving the name 4-methylpentane.

There are two common ways of giving a shortened structural formula for 2-methylpentane:

$$CH_3CH(CH_3)CH_2CH_2CH_3 \quad \text{or} \quad CH_3CHCH_2CH_2CH_3$$
$$\phantom{CH_3CH(CH_3)CH_2CH_2CH_3 \quad \text{or} \quad CH_3CH} | $$
$$\phantom{CH_3CH(CH_3)CH_2CH_2CH_3 \quad \text{or} \quad CH_3CH} CH_3$$

b)

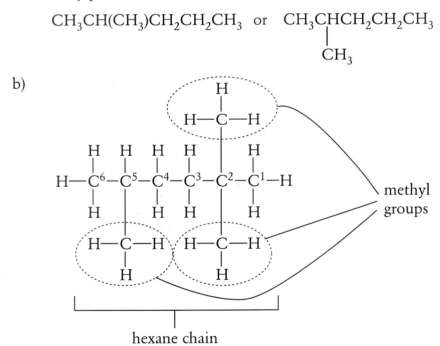

hexane chain

In this example three methyl groups are attached to a hexane chain. Two methyl groups are attached to the second carbon atom and one methyl group to the fifth. The branched chain alkane is therefore given as **2,2,5-trimethylhexane**. Notice that the prefix tri- indicates the presence of *three* methyl groups and that the molecule is numbered from the *right* since this gives smaller numbers than if the molecule's longest chain of carbon atoms was numbered from the left. If the molecule were to be numbered from the left its name would be given as 2,5,5-trimethylhexane.

Two possible ways of giving a shortened structural formula for this particular branched chain alkane are:

$$CH_3CH(CH_3)CH_2CH_2C(CH_3)_2CH_3 \quad \text{or} \quad CH_3CHCH_2CH_2CCH_3$$

with methyl groups:

$$\begin{array}{c} CH_3 \\ | \\ CH_3CHCH_2CH_2CCH_3 \\ | \qquad\qquad | \\ CH_3 \qquad\quad CH_3 \end{array}$$

c)

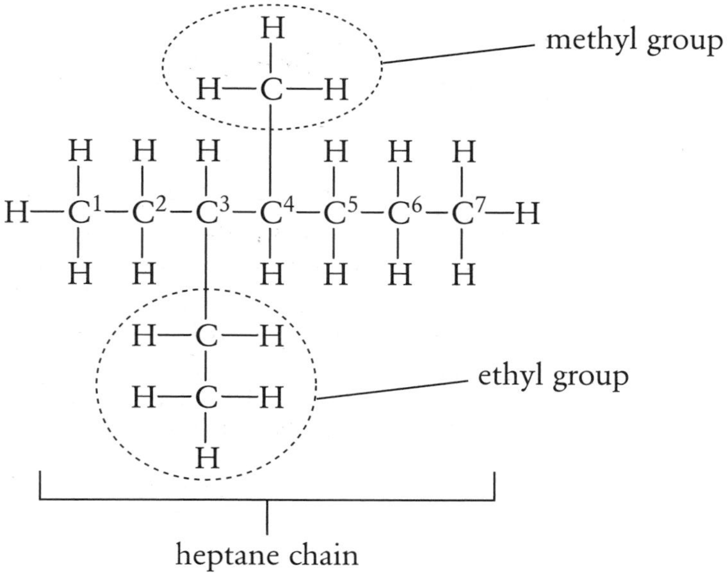

In this example an ethyl and a methyl group are attached to a heptane chain. Numbering of the carbon chain starts from the *left* so that the ethyl group is attached to carbon atom number three and the methyl group to carbon atom number four. Numbering from the right is not correct as this gives larger numbers. Both in terms of their numbered positions and alphabetically, the position of the ethyl group is given *before* the position of the methyl group. The name for this branched chain alkane is therefore **3-ethyl-4-methylheptane**.

Two common ways of writing shortened structural formulae for this branched chain alkane are as follows:

$$CH_3CH_2CH(C_2H_5)CH(CH_3)CH_2CH_2CH_3$$

or

$$\begin{array}{c} CH_3 \\ | \\ CH_3CH_2CHCHCH_2CH_2CH_3 \\ | \\ CH_2CH_3 \end{array}$$

d) It should be noted that, because of free rotation about carbon-to-carbon single covalent bonds, it is legitimate to present structural formulae for branched chain hydrocarbons in which the longest carbon chain is *not* shown in a linear manner. For example, the molecule shown on page 92 is **3-methylpentane**, not 2-ethylbutane.

$$H-\overset{\overset{\displaystyle H}{|}}{\underset{\underset{\displaystyle H}{|}}{C}}-\overset{\overset{\displaystyle H}{|}}{C^3}-\overset{\overset{\displaystyle H}{|}}{\underset{\underset{\displaystyle H}{|}}{C^2}}-\overset{\overset{\displaystyle H}{|}}{\underset{\underset{\displaystyle H}{|}}{C^1}}-H$$

$$H-\overset{}{\underset{}{C^4}}-H$$

$$H-\overset{}{\underset{\underset{\displaystyle H}{|}}{C^5}}-H$$

Questions

7 Give systematic names for each of the following:

(a) $CH_3CH_2CHCH_2CH_3$
    |
    CH_3

(b)

$$H-\overset{\overset{\displaystyle H}{|}}{\underset{\underset{\displaystyle H}{|}}{C}}-\overset{\overset{\displaystyle H}{|}}{\underset{\underset{\displaystyle H}{|}}{C}}-\overset{\overset{\displaystyle H}{|}}{\underset{\underset{\displaystyle H}{|}}{C}}-\overset{\overset{\displaystyle H}{|}}{\underset{\underset{\displaystyle H}{|}}{C}}-\overset{\overset{\displaystyle H-C-H}{|}}{\underset{\underset{\displaystyle H-C-H}{|}}{C}}-\overset{\overset{\displaystyle H}{|}}{\underset{\underset{\displaystyle H}{|}}{C}}-H$$

(c) $CH_3CH_2CH(C_2H_5)C(CH_3)_2CH_2CH_2CH_2CH_3$

Writing structures from names for branched chain alkanes

a) How to write structural formulae for **2,3-dimethylbutane**:

i) The parent alkane is butane, so we draw the numbered carbon skeleton for a molecule of butane with the covalent bonds attached.

$$-\overset{|}{\underset{|}{C^1}}-\overset{|}{\underset{|}{C^2}}-\overset{|}{\underset{|}{C^3}}-\overset{|}{\underset{|}{C^4}}-$$

ii) Attach two methyl groups, with one joined to carbon atom number two and the other to carbon atom number three.

$$H-\overset{\overset{\displaystyle H}{|}}{C}-H$$

$$-\overset{|}{\underset{|}{C^1}}-\overset{|}{\underset{|}{C^2}}-\overset{|}{\underset{|}{C^3}}-\overset{|}{\underset{|}{C^4}}-$$

$$H-\overset{}{\underset{\underset{\displaystyle H}{|}}{C}}-H$$

iii) Attach the remaining hydrogen atoms of the parent butane molecule but remove the numbers from the carbon atoms.

full structural formula for 2,3-dimethylbutane

shortened structural formulae

$$
\begin{array}{c}
\text{H} \\
| \\
\text{H}-\text{C}-\text{H} \\
\text{H} \quad | \quad \text{H} \quad \text{H} \\
| \quad | \quad | \quad | \\
\text{H}-\text{C}-\text{C}-\text{C}-\text{C}-\text{H} \\
| \quad | \quad | \quad | \\
\text{H} \quad \text{H} \quad | \quad \text{H} \\
\text{H}-\text{C}-\text{H} \\
| \\
\text{H}
\end{array}
$$

$$
\begin{array}{c}
\text{CH}_3 \\
| \\
\text{CH}_3\text{CHCHCH}_3 \\
| \\
\text{CH}_3
\end{array}
$$

$$CH_3CH(CH_3)CH(CH_3)CH_3$$

b) How to write structural formulae for **4-ethyl-2-methylhexane**:

i) The parent alkane is hexane for which the numbered carbon skeleton, with covalent bonds attached, is:

$$-\overset{|}{\underset{|}{C^1}}-\overset{|}{\underset{|}{C^2}}-\overset{|}{\underset{|}{C^3}}-\overset{|}{\underset{|}{C^4}}-\overset{|}{\underset{|}{C^5}}-\overset{|}{\underset{|}{C^6}}-$$

ii) An ethyl group is attached to carbon atom number four and a methyl group to carbon atom number two.

$$
\begin{array}{c}
\text{H} \\
| \\
\text{H}-\text{C}-\text{H} \\
| \\
-\overset{|}{\underset{|}{C^1}}-\overset{|}{\underset{|}{C^2}}-\overset{|}{\underset{|}{C^3}}-\overset{|}{\underset{|}{C^4}}-\overset{|}{\underset{|}{C^5}}-\overset{|}{\underset{|}{C^6}}- \\
\text{H}-\text{C}-\text{H} \\
| \\
\text{H}-\text{C}-\text{H} \\
| \\
\text{H}
\end{array}
$$

iii) Attach the remaining hydrogen atoms of the parent hexane molecule but remove the numbers from the carbon atoms.

full structural formula for 4-ethyl-2-methylhexane

shortened structural formulae

$$
\begin{array}{c}
\text{H} \\
| \\
\text{H}-\text{C}-\text{H} \\
\text{H} \quad | \quad \text{H} \quad \text{H} \quad \text{H} \quad \text{H} \\
| \quad | \quad | \quad | \quad | \quad | \\
\text{H}-\text{C}-\text{C}-\text{C}-\text{C}-\text{C}-\text{C}-\text{H} \\
| \quad | \quad | \quad | \quad | \\
\text{H} \quad \text{H} \quad \text{H} \quad | \quad \text{H} \quad \text{H} \\
\text{H}-\text{C}-\text{H} \\
| \\
\text{H}-\text{C}-\text{H} \\
| \\
\text{H}
\end{array}
$$

$$
\begin{array}{c}
\text{CH}_3 \\
| \\
\text{CH}_3\text{CHCH}_2\text{CHCH}_2\text{CH}_3 \\
| \\
\text{CH}_2\text{CH}_3
\end{array}
$$

$$CH_3CH(CH_3)CH_2CH(C_2H_5)CH_2CH_3$$

Using the alkanes

Crude oil and natural gas are made up mainly of alkanes and in both cases they are used mainly as fuels. This is because the alkanes burn, combining with oxygen, to release heat energy. Apart from this they are relatively unreactive. However, they can be decomposed, by heating, to produce a more reactive set of of hydrocarbons called the **alkenes**. This reaction is very important for the petrochemicals industry and is called **cracking**. It is described on page 113.

Alkenes

The **alkenes**, like the alkanes, are a subset of the set of hydrocarbons. They differ from the alkanes in that their molecules each contain a carbon-to-carbon double covalent bond, C=C. As you will see later, the presence of this bond gives the alkenes some properties that are different from those of the alkanes.

Table 8.3 shows the names and formulae for some simple alkenes. There are four points to note:

1 The first part of the alkene names are the same as those of the alkanes.

2 All the alkene names end in **-ene**.

3 There is no alkene with just one carbon atom in its molecules. This is because two carbon atoms are needed to form a C=C bond.

4 In larger alkene molecules, the name contains a number to indicate where the C=C bond is.

Name	Molecular formula and state	Full structural formula	Boiling point °C
ethene	C_2H_4(g)		−102
propene	C_3H_6(g)		−48
but-1-ene	C_4H_8(g)		−6
but-2-ene	C_4H_8(g)		3

Table 8.3 ▶
The simplest alkenes.

94

There are many other alkenes which have even larger molecules. As with the alkanes, the boiling point increases with molecular size – the pentenes being the first alkenes that are liquid at room temperature. It should be noted that, in the case of the butenes, but-1-ene and but-2-ene, both molecules, although different, share the same molecular formula (C_4H_8). This property is called **isomerism** and is dealt with on page 98.

The general formula for the alkenes

From Table 8.3 we can see that, in every alkene molecule, there are twice as many hydrogen atoms as there are carbon atoms. Therefore, if an alkene molecule contains n carbon atoms, the number of hydrogen atoms will be *2n*. This leads to the conclusion that the general molecular formula, or general formula as it is usually called is:

$$C_nH_{2n}$$

Shortened structural formulae

As with the alkanes, shortened structural formulae can also be drawn for the alkenes. Some examples are given in Table 8.4.

Name	Molecular formula	Shortened structural formula
ethene	C_2H_4	$CH_2{=}CH_2$
propene	C_3H_6	$CH_2{=}CHCH_3$
but-1-ene	C_4H_8	$CH_2{=}CHCH_2CH_3$
but-2-ene	C_4H_8	$CH_3CH{=}CHCH_3$
pent-1-ene	C_5H_{10}	$CH_2{=}CHCH_2CH_2CH_3$
pent-2-ene	C_5H_{10}	$CH_3CH{=}CHCH_2CH_3$

Table 8.4 ▲
Shortened structural formulae for some simple alkenes.

Rules for naming alkenes

The rules for naming alkenes are as follows.

1 Look for the *longest* carbon atom chain containing the $C{=}C$ bond to give the name of the parent alkane.

2 Remove the **-ane** ending of the parent alkane and replace it by **-ene**.

3 For chains of four or more carbon atoms, a number must be added in front of the -ene ending to indicate the position of the $C{=}C$ bond.

4 Any number that is used in the name of an alkene refers to the position of the *first* carbon atom to which the double covalent bond is attached.

5 Where two numbers are possible, the *smaller* must be used. e.g.

$$H{-}\overset{\displaystyle H}{\underset{\displaystyle H}{C^1}}{-}\overset{\displaystyle H}{C^2}{=}\overset{\displaystyle H}{C^3}{-}\overset{\displaystyle H}{C^4}{-}\overset{\displaystyle H}{\underset{\displaystyle H}{C^5}}{-}H \qquad \text{is } \textbf{pent-2-ene}, \text{ not pent-3-ene.}$$

6 The positions of any alkyl branches can be indicated using the rules for branched chain alkanes.

Examples of naming alkenes

a)

$$H-\overset{\displaystyle H}{\underset{\displaystyle H}{C^1}}-\overset{\displaystyle H}{\underset{\displaystyle H}{C^2}}-\overset{\displaystyle H}{C^3}=\overset{\displaystyle H}{C^4}-\overset{\displaystyle H}{\underset{\displaystyle H}{C^5}}-\overset{\displaystyle H}{\underset{\displaystyle H}{C^6}}-H$$

There is a chain of six carbon atoms in this alkene, so the parent alkane is hexane and this molecule must therefore be a hex**ene**. When numbered from either end, the first carbon atom in the C=C double covalent bond is number three in the chain. The name of this alkene is therefore **hex-3-ene**.

b)

$$H-\overset{\displaystyle H}{\underset{\displaystyle H}{C^7}}-\overset{\displaystyle H}{\underset{\displaystyle H}{C^6}}-\overset{\displaystyle H}{\underset{\displaystyle H}{C^5}}-\overset{\displaystyle H}{\underset{\displaystyle H}{C^4}}-\overset{\displaystyle H}{C^3}=\overset{\displaystyle H}{C^2}-\overset{\displaystyle H}{\underset{\displaystyle H}{C^1}}-H$$

The parent alkane in this case is heptane because there are seven carbon atoms in the chain. The chain is numbered from the *right* because this gives the *smaller* of the two possible numbers for the first carbon atom in the C=C bond. The name for this alkene is therefore **hept-2-ene** and not hept-5-ene, which would be the case if the chain were to be numbered from the left.

Writing structures from names for alkenes

a) How to write structural formulae for **hex-2-ene**:

   i) Draw the numbered carbon chain for a molecule of hexane.

   $$C^1—C^2—C^3—C^4—C^5—C^6$$

   ii) Insert a carbon-to-carbon double covalent bond, instead of the single one, starting at carbon atom number two.

   $$C^1—C^2{=}C^3—C^4—C^5—C^6$$

   iii) Add hydrogen atoms, by means of single covalent bonds to bring the valency of carbon up to its usual value of four, but omit the numbers on the carbon chain.

   full structural formula for hex-2-ene   shortened structural formula

   $$H-\overset{\displaystyle H}{\underset{\displaystyle H}{C}}-\overset{\displaystyle H}{C}=\overset{\displaystyle H}{C}-\overset{\displaystyle H}{\underset{\displaystyle H}{C}}-\overset{\displaystyle H}{\underset{\displaystyle H}{C}}-\overset{\displaystyle H}{\underset{\displaystyle H}{C}}-H \qquad CH_3CH{=}CHCH_2CH_2CH_3$$

Questions

9 Give systematic names for each of the following:
(a) $CH_2 = CH_2$
(b) $CH_3CH = CHCH_3$
(c) $CH_3CH_2CH_2CH = CHCH_3$

10 Draw the full structural formulae and shortened structural formulae for each of the following:
(a) propene
(b) pent-1-ene
(c) oct-3-ene

b) How to write structural formulae for **hept-3-ene**:
i) Draw the numbered carbon chain for a molecule of heptane.

$$C^1 — C^2 — C^3 — C^4 — C^5 — C^6 — C^7$$

ii) Insert a $C = C$ bond between carbon atoms number three and four.

$$C^1 — C^2 — C^3 = C^4 — C^5 — C^6 — C^7$$

iii) Add hydrogen atoms, by means of single covalent bonds to bring the valency of carbon up to four, but omit the numbers on the carbon chain.

full structural formula for hept-3-ene   shortened structural formula

$$CH_3CH_2CH=CHCH_2CH_2CH_3$$

Cycloalkanes

Like the alkanes and alkenes, the **cycloalkanes** are another subset of the set of hydrocarbons. The alkanes and alkenes show us that carbon atoms can join together to form chains but they can also join to form rings. The cycloalkanes are ring molecules in which *three* or more CH_2 groups are joined by single covalent bonds. The name of a cycloalkane is derived from that of the alkane that has the same number of carbon atoms, but the prefix 'cyclo-' is put in front of it. Table 8.5 shows the names and formulae for the four simplest cycloalkanes.

Name	Molecular formula and state	Full structural formula	Shortened structural formula
cyclopropane	C_3H_6(g)		
cyclobutane	C_4H_8(g)		
cyclopentane	C_5H_{10}(l)		
cyclohexane	C_6H_{12}(l)		

Table 8.5 ▲
The simplest cycloalkanes.

11 Draw the full structural formulae and molecular formulae for each of the following:
  (a) cyclobutane
  (b) cycloheptane
  (c) cyclooctane

The general formula for the cycloalkanes

As with the alkenes, we can see that every cycloalkane molecule contains twice as many hydrogen atoms as carbon atoms and therefore has the same general formula. If the number of carbon atoms in a molecule is n then the number of hydrogen atoms present must be $2n$. The general formula for the cycloalkanes is therefore:

$$C_nH_{2n}$$

Since their general formulae are the same, there are many examples where alkenes and cycloalkanes have the same molecular formula.

Using the cycloalkanes

Only relatively small amounts of cycloalkanes occur naturally in crude oil, but they are important as part of the mixture sold as 'unleaded petrol'. As part of the refining of crude oil to produce unleaded petrol, straight chain alkanes in naphtha are subjected to a process called 'reforming'. This causes many of the molecules to break up and then form cycloalkanes, branched chain alkanes and other molecules that provide a better quality of petrol than alkanes alone would produce. Cyclohexane is also used in the plastics industry for the production of several types of nylon that are widely used in the clothing industry. Other examples of the use of nylon include carpet piles, brushes and fishing nets and lines.

Homologous series

A **homologous series** is a group of compounds with similar chemical properties and which can be represented by a general formula. Their physical properties, such as boiling points, show a gradual change with increasing molecular size. Boiling points of members of a homologous series increase with increasing molecular size because the forces of attraction between the molecules become larger and more energy is needed to separate them. The alkanes, alkenes and the cycloalkanes are all examples of homologous series.

In summary we can say that a homologous series is a group of compounds that:

♦ can be represented by a general formula
♦ have similar chemical properties
♦ show a gradual change in physical properties, such as boiling points.

Isomers

In this chapter we have seen that different compounds can have the same molecular formulae. Examples include 2-methylpropane and butane, which share the same molecular formula C_4H_{10} (see Figure 8.1 on page 89). Because they have different structural formulae, but have

12 Draw full structural formulae for three of the isomers with molecular formula C_6H_{14} and give their systematic names.

13 Draw shortened structural formulae for the straight chain alkene isomers with molecular formula C_7H_{14} and give their systematic names.

14 Draw full structural formulae for two isomers that have molecular formula C_4H_8 and belong to two different homologous series. Give systematic names for the two isomers.

the same molecular formulae, they are called **isomers**. Full structural formulae for these isomers are shown below.

H   H   H   H
| | | |
H—C—C—C—C—H
| | | |
H   H   H   H

butane

H   H   H
| | |
H—C—C—C—H
| | |
H   H
|
H—C—H
|
H

2-methylpropane

The number of isomers with a particular molecular formula increases with the number of atoms present. For example there are three isomers with molecular formula C_5H_{12}:

H   H   H   H   H
| | | | |
H—C—C—C—C—C—H
| | | | |
H   H   H   H   H

pentane

H   H   H   H
| | | |
H—C—C—C—C—H
| | | |
H   H   H   H
|
H—C—H
|
H

2-methylbutane

H
|
H—C—H
|
H   H   H
| | |
H—C—C—C—H
| | |
H   H   H
|
H—C—H
|
H

2,2 dimethylpropane

The only alkanes which have no isomers are the three simplest ones, namely methane (CH_4), ethane (C_2H_6) and propane (C_3H_8).

All of the alkenes, with the exceptions of ethene (C_2H_4) and propene (C_3H_6), have other straight chain alkene isomers. For example, there are two straight chain alkene isomers with molecular formula C_4H_8.

H   H   H   H
| | | /
H—C—C—C=C
| | \
H   H   H

$CH_3CH_2CH{=}CH_2$
but–1–ene

H   H   H   H
| | | |
H—C—C=C—C—H
| |
H   H

$CH_3CH{=}CHCH_3$
but–2–ene

Earlier in this section we saw that both the alkenes and the cycloalkanes have the same general formula of C_nH_{2n}. This means that for every cycloalkane there is an alkene with the same molecular formula but with a different structural formula. Two such isomers, propene and cyclopropane, which both have the molecular formula C_3H_6, are shown on page 100.

cyclopropane   propene

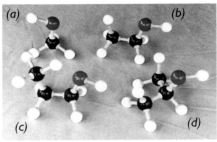

Figure 8.2 ▲
*Molecular models of a) methanol,
b) ethanol, c) propan-l-ol and
d) propan-2-ol.*

Alkanols

The **alkanols** form a homologous series that is very similar in structure to the alkanes. The only difference is that one of the hydrogen atoms in an alkane has been replaced by a **hydroxyl group**, –OH.

Table 8.6 shows the names and formulae for some simple alkanols. When naming the alkanols, the '-e' ending of the parent alkane is removed and '-ol' is added instead. As with the alkenes, a number is inserted to show the carbon atom to which the –OH group is attached if the molecule is large enough to have isomers.

Name	Molecular formula and state	Full structural formula	Shortened structural formula	Boiling point °C
methanol	$CH_3OH(l)$		CH_3OH	65
ethanol	$C_2H_5OH(l)$		CH_3CH_2OH	79
propan-1-ol	$C_3H_7OH(l)$		$CH_3CH_2CH_2OH$	97
propan-2-ol	$C_3H_7OH(l)$		$CH_2CHOHCH_2OH$	82

Table 8.6 ▲
The simplest alkanols.

As with all homologous series, the boiling points of the alkanols increase as the molecules become larger. However, unlike hydrocarbons, all of the simple alkanols are liquid at room temperature because there are stronger forces of attraction between the molecules.

There are no isomeric alkanols with molecular formulae CH_3OH (methanol) and C_2H_5OH (ethanol), but when there are three or more carbon atoms present then different alkanols share the same molecular formula. Propan-1-ol and propan-2-ol are the first examples of this.

The general formula for the alkanols

The alkanols are simply alkanes that have an extra oxygen atom. As a result, since the general formula for the alkanes is C_nH_{2n+2}, the general formula for the alkanols can be written as $\mathbf{C_nH_{2n+2}O}$. However, since the presence of the –OH group gives the alkanols special chemical properties, it is usually included in the general formula as shown below.

$$C_nH_{2n+1}OH$$

Rules for naming alkanols

1 Look for the *longest* carbon atom chain containing the –OH group to give the name of the parent alkane.

2 Remove the **-e** ending of the parent alkane and add **-ol**.

3 For chains of three or more carbon atoms, a number must be added in front of -ol to indicate the position of the –OH group.

4 The carbon chain is numbered from the end that gives the *smaller* number to the –OH group.

e.g.

$$
\begin{array}{ccccc}
 & H & H & H & H \\
 & | & | & | & | \\
H- & C^4 & -C^3 & -C^2 & -C^1-H \\
 & | & | & | & | \\
 & H & H & OH & H
\end{array}
$$

is **butan-2-ol**, not butan-3-ol.

(Note that in this structure the –OH group is not given in full structural form as – O–H. This is quite normal practice.)

Examples of naming alkanols

a)

$$
\begin{array}{ccccccc}
 & H & H & H & H & H & H \\
 & | & | & | & | & | & | \\
H- & C^6 & -C^5 & -C^4 & -C^3 & -C^2 & -C^1-H \\
 & | & | & | & | & | & | \\
 & H & H & H & H & H & OH
\end{array}
$$

The chain of six carbon atoms indicates that the parent alkane of this alkanol is hexane and that this molecule must therefore be a hexan**ol**. The chain is numbered from the *right* because this gives the smaller of the two possible numbers for the position of the –OH group. The name of the alkanol is therefore **hexan-1-ol**, not hexan-6-ol.

b)

$$
\begin{array}{ccccccccc}
 & H & H & H & H & H & H & H & H \\
 & | & | & | & | & | & | & | & | \\
H- & C^1 & -C^2 & -C^3 & -C^4 & -C^5 & -C^6 & -C^7 & -C^8-H \\
 & | & | & | & | & | & | & | & | \\
 & H & H & OH & H & H & H & H & H
\end{array}
$$

The parent alkane in this case is octane because there are eight carbon atoms in the chain. The chain is numbered from the *left* because this gives the *smaller* of the two possible numbers for the position of the

Figure 8.3 ▲
The alcohol referred to in the list of ingredients is ethanol.

–OH group. The name for this alkanol is therefore **octan-3-ol**, not octan-6-ol, which would be the name if the chain was numbered from the right.

Using the alkanols

There is a wide variety of uses for alkanols. Methanol is used as a feedstock for making some plastics and is the fuel used in Indy Car racing. Ethanol, or ethyl alcohol as it is traditionally known, is the most important alkanol. It is present in all alcoholic drinks, but is also found in many cosmetics (see Figure 8.3) and is widely used as a solvent. Ethanol can also be used successfully as a liquid fuel in place of petrol.

Alkanoic acids

The **alkanoic acids** form a homologous series in which a $-CH_3$ group in an alkane has been replaced by a **carboxyl group** –COOH. The full structural formula for the carboxyl group is as follows:

$$\begin{array}{c} \quad\quad O \\ \quad\quad \| \\ -C \\ \quad\quad \backslash \\ \quad\quad O-H \end{array}$$

Table 8.7 shows the names and formulae for some simple alkanoic acids. When naming the alkanoic acids, the '-e' ending of the parent alkane is removed and replaced by '-oic acid'.

Question

15 Draw the full structural formulae, shortened structural formulae and molecular formulae for each of the following:
(a) propan-2-ol
(b) pentan-1-ol
(c) heptan-3-ol

Name	Molecular formula and state	Full structural formula	Shortened structural formula	Boiling point °C
methanoic acid	HCOOH(l)		HCOOH	101
ethanoic acid	$CH_3COOH(l)$		CH_3COOH	118
propanoic acid	$C_2H_5COOH(l)$		$CH_3CH_2CH_2COOH$	141
butanoic acid	$C_3H_7COOH(l)$		$CH_3CH_2CH_2CH_2COOH$	164

Table 8.7 ▲
The simplest alkanoic acids.

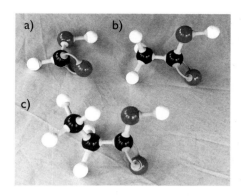

Figure 8.4 ▲
*Molecular models of a) methanoic acid,
b) ethanoic acid and c) propanoic acid.*

Like the alkanols, the simplest alkanoic acids are liquid at room temperature and their boiling points increase as the molecules become larger. The forces of attraction between alkanoic acid molecules are even greater than those between the corresponding alkanol molecules, so the boiling points of the alkanoic acids are higher.

The general formula for the alkanoic acids
From Table 8.7 it can be worked out that the general formula for the alkanoic acids is $C_nH_{2n+1}COOH$, where n = 0, 1, 2, etc. When n = 0 the molecular formula is HCOOH (methanoic acid).

Rules for naming alkanoic acids

1 Count the number of atoms in the carbon chain to give the name of the parent alkane.

2 Remove the **-e** ending of the parent alkanol and replace it by **-oic acid**.

Using the alkanoic acids
Both methanoic acid and ethanoic acid are important in the chemical industry for the manufacture of other products, but both have simple everyday uses. Methanoic acid is sold as a remover of limescale in kettles, which is caused by hard water. Ethanoic acid, which is more commonly recognised by its older name of acetic acid, is used as a food additive (E260) and as a dilute solution is sold as vinegar.

Esters

Esters are compounds that are formed when an alkanol reacts with an alkanoic acid. Water is also produced during the reaction. The general word equation for ester formation is therefore:

$$\text{alkanol} + \text{alkanoic acid} \rightarrow \text{ester} + \text{water}$$

Rules for naming esters

1 The names for esters are based on the alkanol and alkanoic acid from which they are made.

2 Part of the name comes from the alkanol and part from the alkanoic acid.

3 Esters names are therefore usually written as two words of the type *alkyl alkanoate*.

4 The *alkyl* part comes from the alkanol and the *alkanoate* part from the alkanoic acid.

   e.g. Methanol produces *methyl* esters.
   Ethanol produces *ethyl* esters.
   Propan-1-ol produces *propyl* esters.

Questions

16 Draw the full structural formulae and shortened structural formulae for each of the following:
   (a) hexanoic acid
   (b) propanoic acid

17 Name the following alkanoic acids:
   (a) $CH_3CH_2CH_2COOH$
   (b) $CH_3CH_2CH_2CH_2CH_2CH_2COOH$

e.g. Methanoic acid produces *methanoate* esters.
Ethanoic acid produces *ethanoate* esters.
Propanoic acid produces *propanoate* esters.

Some word equations showing ester formation

a) methanol + ethanoic acid → methyl ethanoate + water

b) ethanol + methanoic acid → ethyl methanoate + water

Ester formation using structural formulae

The formation of an ester involves a reaction between the hydroxyl group in an alkanol and the carboxyl group in an alkanoic acid. The molecules join to form an ester molecule, with a water molecule formed as the joining takes place. The 'general reaction' is as follows:

| alkanoic acid | alkanol | ester | water |

represents the rest of the alkanoic acid.

represents the rest of the alkanol molecule.

The group of atoms is called the **ester group** or **ester link**.

Shortened structural formulae may also be used to show the formation of an ester:

| alkanoic acid | alkanol | ester | water |

The ester group in its shortened form therefore appears as –COO–.

It is quite acceptable to draw the formula of the alkanol *before* that of the alkanoic acid and this will be given in one of the examples that follow.

Examples of equations resulting in ester formation

a) i) word equation showing the alkanoic acid first:

ethanoic acid + methanol → methyl ethanoate + water

ii) equation showing full structural formulae:

H   O   H   O
| ‖ | ‖
H—C—C   H—C—C   H
| O—H + H—O—C—H   → | |
H | H   O—C—H + H₂O
                    H |
                                    H

ethanoic acid   methanol   methyl ethanoate   water

iii) equation showing shortened structural formulae:

$$CH_3COOH + CH_3OH \rightarrow CH_3COOCH_3 + H_2O$$

ethanoic acid   methanol   methyl ethanoate   water

b) i) word equation showing the alkanol first:

ethanol + methanoic acid → ethyl methanoate + water

ii) equation showing full structural formulae:

H H   O   H H   O
| | ‖ | | ‖
H—C—C—O—H + H—O—C—H   →   H—C—C—O—C—H + H₂O
| | | |
H H   H H

ethanol   methanoic acid   ethyl methanoate   water

iii) equation showing shortened structural formulae:

$$CH_3CH_2OH + HCOOH \rightarrow CH_3CH_2OOCH + H_2O$$

ethanol   methanoic acid   ethyl methanoate   water

Breaking down esters

Esters can be broken down quite easily by reaction with water into the corresponding alkanol and alkanoic acid. For example:

a) Methyl esters break down to give methanol.
   Ethyl esters break down to give ethanol.
   Propyl esters break down to give propan-1-ol.

b) Methanoate esters break down to give methanoic acid.
   Ethanoate esters break down to give ethanoic acid.
   Propanoate esters break down to give propanoic acid.

Examples of equations showing the breakdown of esters

a) i) word equation:

ethyl ethanoate + water → ethanoic acid + ethanol

ii) equation showing full structural formulae (acid part first):

ethyl ethanoate   water   ethanoic acid   ethanol

iii) equation showing shortened structural formulae:

$$CH_3COOCH_2CH_3 + H_2O \rightarrow CH_3COOH + CH_3CH_2OH$$

ethyl ethanoate   water   ethanoic acid   ethanol

b) i) word equation:

methyl propanoate + water → methanol + propanoic acid

ii) equation showing full structural formulae (alkanol part first):

methyl propanoate   water   methanol   propanoic acid

iii) equation showing shortened structural formulae:

$$CH_3CH_2COOCH_3 + H_2O \rightarrow CH_3OH + CH_3CH_2COOH$$

methyl propanoate   water   methanol   propanoic acid

Rules for naming esters

1 Decide which alkanol was used to form the ester.

2 Remove the **-anol** ending of the alkanol and add **-yl**.
   e.g. eth**anol** becomes eth**yl**.

3 Decide which alkanoic acid was used to form the ester.

4 Remove the **-ic** ending of the alkanoic acid and add **-ate**.
   e.g. methano**ic** acid becomes methano**ate**.

5 Put the two words together to give the name of the ester with the
   alkyl group first. e.g. eth**anol** and methano**ic** acid react to produce
   the ester eth**yl** methano**ate**.

Using the esters

Simple esters are sweet-smelling liquids and are widely used as fruit flavourings (see Table 8.8).

Ester name	Flavour
ethyl methanoate	raspberries
pentyl butanoate	apricots
octyl ethanoate	oranges
pentyl ethanoate	pears

Table 8.8 ▲
Some esters and their flavours.

Industrially and cosmetically, simple esters are important solvents, for example in paint and varnish. Nail varnish and nail varnish remover both contain esters. A variety of esters is used in making perfumes.

Questions

18 Name the ester produced when the following react:

  (a) propan-1-ol and methanoic acid
  (b) methanol and pentanoic acid.

19 Name the alkanol and alkanoic acid that would react to produce the following esters:

  (a) methyl propanoate
  (b) ethyl hexanoate.

20 (a) Draw full structural formulae for the esters referred to in question 19.

  (b) Draw shortened structural formulae for these same esters.

Chapter 8 Study Questions

In questions 1–4 choose the correct word(s) **from the following list** to complete the sentence.

   alkanes, alkenes, cycloalkanes, alkanols, esters, alkanoic acids, isomers, homologous series.

1 The _____ contain a carbon-to-carbon double covalent bond, C=C.

2 The simplest member of the _____ contains three carbon atoms in each molecule.

3 But-1-ene and cyclobutane are examples of _____.

4 The carboxyl group is present in _____.

5 Which of the following is a full structural formula of the carbon compound formed when ethanoic acid reacts with methanol?

A

$$H-\overset{\overset{\displaystyle H}{|}}{\underset{\underset{\displaystyle H}{|}}{C}}-\overset{\overset{\displaystyle H}{|}}{\underset{\underset{\displaystyle H}{|}}{C}}-O-\overset{\overset{\displaystyle O}{||}}{C}-H$$

B

$$H-O-\overset{\overset{\displaystyle O}{||}}{C}-\overset{\overset{\displaystyle H}{|}}{\underset{\underset{\displaystyle H}{|}}{C}}-\overset{\overset{\displaystyle H}{|}}{\underset{\underset{\displaystyle H}{|}}{C}}-H$$

C

$$H-\overset{\overset{\displaystyle H}{|}}{\underset{\underset{\displaystyle H}{|}}{C}}-\overset{\overset{\displaystyle O}{||}}{C}-O-\overset{\overset{\displaystyle H}{|}}{\underset{\underset{\displaystyle H}{|}}{C}}-H$$

D

$$H-\overset{\overset{\displaystyle H}{|}}{\underset{\underset{\displaystyle H}{|}}{C}}-\overset{\overset{\displaystyle O}{||}}{C}-\overset{\overset{\displaystyle H}{|}}{\underset{\underset{\displaystyle H}{|}}{C}}-O-H$$

6 Which of the following could be the molecular formula of an alkane?

A $C_{12}H_{24}$
B $C_{10}H_{22}$
C $C_{14}H_{26}$
D $C_{9}H_{18}$

7 Which one of the following represents a compound that is different from the others?

A

$$CH_3-CH_2-CH_2-\overset{\overset{\displaystyle CH_3}{|}}{C}=CH_2$$

B

$$CH_3-CH_2-CH=\overset{\overset{\displaystyle CH_3}{|}}{C}-CH_3$$

C

$$CH_3-CH_2-CH_2-\overset{\overset{\displaystyle CH_2}{||}}{C}-CH_3$$

D

$$CH_3-\overset{\overset{\displaystyle CH_2}{||}}{C}-CH_2-CH_2-CH_3$$

8 Draw full structural formulae for each of the following:

   (a) propane
   (b) but-2-ene
   (c) cyclopentane

9 Copy and complete the following:

   The members of a homologous series can be represented by a _____ formula, have similar _____ properties and show a gradual change in _____ properties.

10 Give names for each of the following:

   (a)
   $$CH_3CH_2CH_2\overset{\overset{\displaystyle}{}}{\underset{\underset{\displaystyle CH_3}{|}}{C}}HCH_3$$

   (b) $CH_3CHOHCH_3$
   (c) $CH_3CH_2CH_2COOH$

11 Name the esters that are formed by reaction between:

   (a) methanol and methanoic acid
   (b) pentanoic acid and ethanol

12 Give the name of the compound shown below:

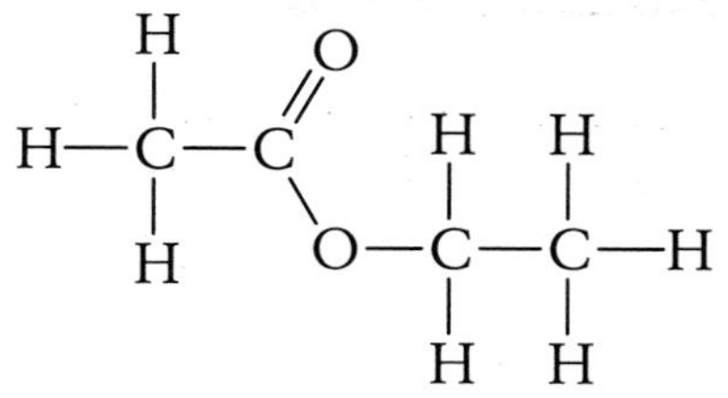

13 Give the general formula for the cycloalkanes.

14 The table below shows the boiling points for some alkanoic acids.

Alkanoic acid	Formula mass	Boiling point °C
methanoic acid	46	101
ethanoic acid	60	118
propanoic acid	74	141
butanoic acid	88	164

(a) Present this information as a bar graph.
(b) Predict the boiling point of the next acid in the series

9 Reactions of Carbon Compounds

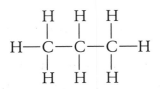

a) propane

b) propene

c) cyclopropane

Figure 9.1 ▲

Addition reactions

Hydrocarbons have some reactions in common. For example, the alkanes, alkenes and cycloalkanes all take part in combustion reactions – in other words they all burn. However, in many ways the alkenes are much more reactive than either the alkanes or the cycloalkanes, but why should this be? The reason lies in a fundamental difference in the types of covalent bonds that are present.

Look at Figure 9.1, which shows full structural formulae for propane, propene and cyclopropane. The reason for propene, an alkene, being so much more reactive than either propane or cyclopropane is that it contains a carbon-to-carbon *double* covalent bond, C=C, which is called a **functional group**. This is a group of atoms with characteristic chemical activity.

The C=C bond is very reactive and easily breaks to give a carbon-to-carbon single covalent bond, C–C, allowing other atoms to join on to each of the carbon atoms in the bond. For example, when heated in the presence of a catalyst, propene can react with hydrogen to produce propane (see Figure 9.2).

propene   hydrogen   propane

The propane molecules that have been formed in this reaction are 'full up' or **saturated** with hydrogen atoms – no more hydrogen atoms can be 'added on'. All alkanes and cycloalkanes are **saturated** because they contain only single carbon-to-carbon covalent bonds so no more atoms can be 'added on'.

Molecules, like propene, are said to be **unsaturated** because they contain one or more carbon-to-carbon double covalent bonds. The presence of this type of bond enables atoms of other elements to 'add on' to the molecule. All of the alkenes are **unsaturated** because the contain a C=C bond so, like propene, they can react with hydrogen to produce the corresponding alkane, which is saturated.

alkene (unsaturated) + **hydrogen** → **alkane** (saturated)

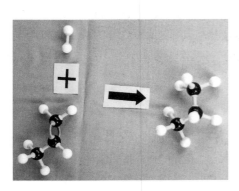

Figure 9.2 ▲
Propene and hydrogen can react to give propane.

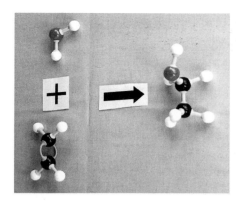

Figure 9.3 ▲
Ethene and water can react to give ethanol.

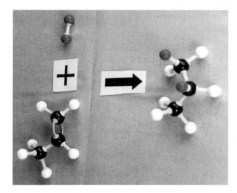

Figure 9.4 ▲
Propene and bromine react to give 1.2-dibromopropane.

The reaction between an alkene and hydrogen is an example of an **addition reaction**. It is given this name because it is one in which two molecules join to produce a single larger molecule and nothing else. All of the atoms that were in the two reacting molecules are found in the one product molecule.

Many elements and compounds can take part in addition reactions with compounds, like the alkenes, that contain carbon-to-carbon double covalent bonds. Two important addition reactions involve the addition of water and bromine to unsaturated compounds like the alkenes.

The reaction between an alkene and water does not take place readily, but it provides an important industrial way of producing alkanols. For example, using heat, pressure and a catalyst, ethene and water react to produce ethanol (see Figure 9.3).

$$\underset{\text{ethene}}{\overset{\displaystyle H \atop \displaystyle H}{\underset{\displaystyle H \atop \displaystyle H}{C=C}}} \quad + \quad \underset{\text{water}}{H-O-H} \quad \longrightarrow \quad \underset{\text{ethanol}}{H-\overset{H}{\underset{H}{C}}-\overset{H}{\underset{O-H}{C}}-H}$$

The general word equation for this reaction is as follows:

$$\textbf{alkene + water} \rightarrow \textbf{alkanol}$$

Unlike the addition of hydrogen and water, bromine adds on readily to a C=C bond without the need for a catalyst or any special conditions. Because the reaction is also accompanied by a clear colour change, chemists use the reaction with bromine as a test for unsaturation.

When shaken with unsaturated compounds, such as the alkenes, bromine solution is rapidly decolourised. For example, propene and bromine react to produce a colourless dibromoalkane called 1,2-dibromopropane (see Figure 9.4).

$$\underset{\substack{\text{propene}\\ \text{(colourless)}}}{H-\overset{H}{\underset{H}{C}}-\overset{H}{C}=\overset{H}{\underset{H}{C}}} \quad + \quad \underset{\substack{\text{bromine}\\ \text{yellow/orange/red}}}{Br-Br} \quad \longrightarrow \quad \underset{\substack{\text{1,2-dibromopropane}\\ \text{(colourless)}}}{H-\overset{H}{\underset{H}{C}}-\overset{H}{\underset{Br}{C}}-\overset{H}{\underset{Br}{C}}-H}$$

The colour of a solution of bromine in water can vary considerably. When very dilute it appears yellow, but as the concentration increases the colour changes, first to orange and then to red.

The general word equation for the reaction between an alkene and bromine is:

alkene + bromine → dibromoalkane

Prescribed Practical Activity

Safety note

Because of the toxic nature of bromine, this experiment is best carried out in a fume chamber.

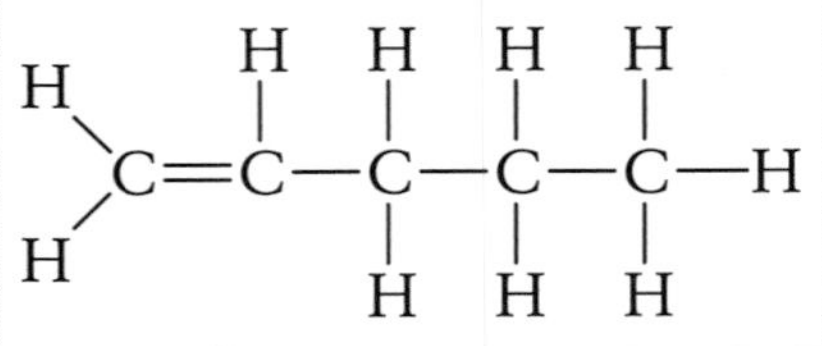

Figure 9.5 ▲

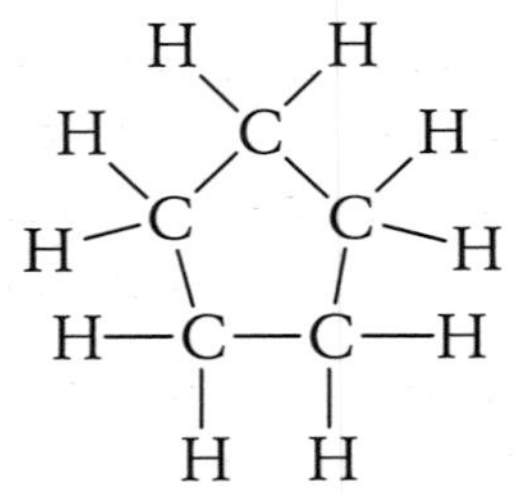

Figure 9.6 ▲

Testing for unsaturation

- The aim of this experiment is to test several liquid hydrocarbons for unsaturation using bromine solution and, based on the results, to draw possible structural formulae for them.
- Add a small volume of a liquid hydrocarbon to a test tube.
- Add a few drops of bromine solution to the hydrocarbon.
- Carefully shake the contents of the test tube and record your observations.
- Decolourisation indicates the presence of at least one C=C bond.
- No decolourisation indicates that there are no C=C bonds present. All carbon-to-carbon covalent bonds present are single, C–C.
- If a hydrocarbon with molecular formula C_5H_{10} *did* decolourise bromine solution, it could be pent-l-ene, the full structural formula of which is shown in Figuire 9.5
- If a hydrocarbon with molecular formula C_5H_{10} did *not* decolourise bromine solution then it could be cyclopentane, the full structural formula of which is shown in Figure 9.6.

Questions

1 (a) Name the compound formed when ethene and hydrogen react by addition.
  (b) Write an equation for the reaction of propene with hydrogen using molecular formulae.

2 (a) Write an equation using full structural formulae showing propene and water reacting to form propan-2-ol.
  (b) Write an equation for the reaction between ethene and water to produce ethanol using molecular formulae.

3 (a) Write an equation using full structural formulae showing ethene reacting with bromine.
  (b) Write an equation for the reaction of propene with bromine using molecular formulae.

Figure 9.7 ▲
A catalytic cracking plant at the BP/Amoco refinery in Grangemouth.

Cracking

The fractional distillation of crude oil does not produce enough gasoline and naphtha to meet the demand for petrol. Because of this a lot of the larger hydrocarbon molecules in crude oil (mainly long chain alkanes) are broken into smaller, more useful molecules by a reaction called **cracking**. Originally cracking was simply used to obtain more petrol from crude oil, but now it is used to provide a wide variety of important substances for use in the petrochemicals industry.

Cracking can take place simply by heating, but the presence of aluminium oxide or a silicate as catalyst enables the reaction to take place at a lower temperature. This saves energy and therefore reduces costs. **Catalytic cracking** is an important chemical reaction at an oil refinery and is used following the physical process of fractional distillation. A photograph of a catalytic cracking plant is shown in Figure 9.7.

The cracking of large alkane molecules usually produces a mixture of smaller alkanes and alkenes, though other products can be obtained as well. For example, the cracking of decane can produce a mixture of octane and ethene.

$$C_{10}H_{22} \rightarrow C_8H_{18} + C_2H_4$$

The usual reaction that occurs during cracking can be summarised as follows:

larger alkane → smaller alkane + smaller alkene

It is not possible for two smaller alkane molecules to be made from one larger alkane molecule because there are insufficient hydrogen atoms present.

Questions

4 Write an equation using molecular formulae to show the cracking of octane to give propene and another alkane.

5 What is the main difference between fractional distillation and cracking?

During cracking some hydrocarbon molecules are broken down completely to carbon and hydrogen. Hydrogen is a valuable by-product, but carbon coats the catalyst reducing its efficiency. Fortunately, the carbon coating on the catalyst is easily burned off and does not hinder the continuous cracking reaction.

The alkenes that are produced by cracking, such as ethene and propene, are very important in the petrochemicals industry. They are used as starting materials for a range of products from plastics to 'antifreeze'.

Prescribed Practical Activity

Safety note

Because of the toxic nature of bromine, this experiment is best carried out in a fume cupboard.

Avoid 'suck-back'.

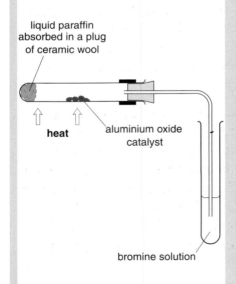

liquid paraffin absorbed in a plug of ceramic wool

heat

aluminium oxide catalyst

bromine solution

Figure 9.8 ▲

Cracking

- The aim of the experiment is to crack liquid paraffin (a mixture of long chain alkanes) and to show that some of the products are unsaturated.
- The apparatus is set up as in Figure 9.8, then the catalyst is heated strongly for about 10 seconds.
- The flame is then directed very briefly at the nearest end of the plug for a second or two to vaporise some paraffin, then back to the catalyst again.
- The process of alternate heating is repeated as required in order to maintain a flow of 'cracked gas'.
- When the bromine solution has been decolourised, a special procedure *must* be followed when heating is to stop.
- The delivery tube must be removed from the solution *before* heating stops so as to prevent 'suck-back'.
- 'Suck-back' happens when pressure inside the apparatus drops and air pressure forces the cold solution up the delivery tube and into the still very hot test tube.
- One constituent of liquid paraffin is eicosane, $C_{20}H_{42}$. When cracked, one possible pair of products is heptadecane, a liquid ($C_{17}H_{36}$), and the unsaturated gas propene (C_3H_6).

$$C_{20}H_{42} \rightarrow C_{17}H_{36} + C_3H_6$$

Ethanol

Ethanol is one of the most versatile of carbon compounds: as was mentioned on page 102, it is the 'alcohol' in all alcoholic drinks, it is found in many cosmetic preparations, it is a widely used solvent and it is a liquid fuel replacement for petrol. There are two ways in which it is made – **fermentation** and **catalytic hydration**.

Fermentation

The word fermentation is derived from the Latin word *fermere*, which means 'to boil'. Sugars are said to ferment when, in the presence of enzymes in yeast cells, they break down exothermically producing ethanol and carbon dioxide gas. This gas causes bubbling and frothing which makes the mixture look as though it is boiling – hence the name, 'fermentation'.

Figure 9.9 ▲
Grapes, from which wine can be made by fermentation, contain the sugar glucose.

Yeast contains many enzymes and these act as catalysts for the fermentation of a wide variety of sugars, such as maltose, sucrose, fructose and glucose. In each case the eventual products are ethanol and carbon dioxide. Most wine is made from grapes (see Figure 9.9). These contain the sugar, glucose, inside them and they have yeast cells on the outside. As a result, when grapes are crushed fermentation starts. The balanced equation for the reaction taking place is:

$$C_6H_{12}O_6 \rightarrow 2C_2H_5OH + 2CO_2$$

glucose   ethanol   carbon dioxide

Just as alcoholic drinks, if taken in excess, can have damaging effects on body and mind, so the ethanol produced has a harmful effect on the yeast cells. These die when the ethanol concentration reaches about 12% and fermentation ceases. You will find that this closely matches the alcohol content of many wines.

Ethanol has a boiling point of 79°C, compared to water's boiling point of 100°C. As a result, pure ethanol can be obtained by distillation of a mixture of ethanol and water (see Figure 9.10). In practice, the separation is quite difficult because of attraction between the ethanol and water molecules and, as an examination of labels will reveal, most of the high-alcohol content drinks contain about 40% ethanol. An alcoholic drink that is obtained by the distillation of wines made from grapes is brandy. Brandy and other high-alcohol drinks like whiskey are called 'spirits'.

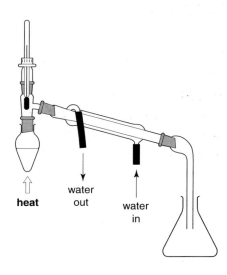

heat   water out   water in

Figure 9.10 ▲
Apparatus for distillation.

In some countries, such as Brazil, ethanol produced by fermentation of sucrose from sugar cane is already being used as a fuel for motor vehicles. A mixture of about 10% ethanol and 90% petrol can be used without any modification to the engine. The energy costs involved in distillation to obtain pure ethanol are high, but the process will become more economical as supplies of crude oil decrease.

Unlike petrol from the non-renewable energy source of crude oil, ethanol is obtained from the **renewable energy source** of sugar cane. This means that it can be renewed (by growing more sugar cane) and will not run out in the forseeable future.

Eventually more countries may use ethanol as a petrol additive, or substitute. In many homes at Christmas time we make use of the fact that ethanol is a liquid fuel when we pour warm brandy over the traditional pudding, light it, then watch it burn with a clean blue flame.

Catalytic hydration

All of the ethanol that was required for alcoholic drinks and for industry used to be made by fermentation. However, the demand for ethanol by industry is now so great that fermentation alone cannot satisfy demand. As a result, chemists have devised a new reaction for its production called **catalytic hydration.**

The reaction between alkenes and water was referred to on page 110. Because it is a special kind of addition reaction in which water molecules add on to a carbon-to-carbon double covalent bond in the presence of a catalyst, it is called catalytic hydration. With fairly high pressure and temperature, ethene and water molecules join to produce ethanol.

ethene   water   ethanol

Using molecular formulae the reaction is:

$$C_2H_4 + H_2O \rightarrow C_2H_5OH$$

Dehydration

Although of lesser importance industrially, it is interesting to note that the hydration reaction above can be reversed and is then called **dehydration**. This is the removal of water from a compound. Ethanol can be easily dehydrated in the laboratory by passing the vapour over hot aluminium oxide, which acts as a catalyst (see Figure 9.11). The reaction taking place is:

ethanol   ethene   water

Questions

6 Why can ethanol be separated from water by distillation?

7 Why is ethanol said to be a 'renewable fuel'?

8 Propan-2-ol is made by the catalytic hydration of propene. Write an equation for this reaction using shortened structural formulae.

Many alkanols can be dehydrated using this same method, producing one or more alkenes, depending on the structure of the alkanol molecule.

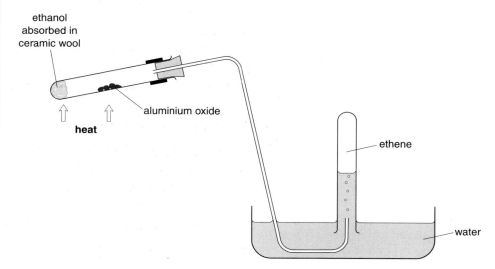

Figure 9.11 ▲
Ethanol can be dehydrated to give ethene.

Making and breaking esters

Making esters – a condensation reaction

The reaction between an alkanol and an alkanoic acid to produce an ester and water is **reversible**, meaning that it proceeds in both directions. Alkanol and alkanoic acid react to form ester and water, but at the same time ester and water react to reform alkanol and alkanoic acid. A reversible reaction is shown by using a special kind of double-headed arrow, $\rightleftharpoons$. The reaction between an alkanol and an alkanoic acid to make an ester and water is therefore shown as:

$$\textbf{alkanol} + \textbf{alkanoic acid} \rightleftharpoons \textbf{ester} + \textbf{water}$$

Esters can be made by mixing small equal volumes (about 2 cm³) of an alkanol and an alkanoic acid in a test tube. To this mixture a few drops of concentrated sulphuric acid are added as catalyst. The mixture is gently shaken, then placed in a hot water bath, as shown in Figure 9.12, for about ten minutes.

Most esters are both less dense than water and insoluble in it, so the ester can be separated from the other, water-soluble compounds present by pouring the mixture into water. The ester then floats on the top of the water (see Figure 9.13). If sodium hydrogencarbonate solution is used instead of water, then the unreacted acids are neutralised and the strong smell of the alkanoic acids is completely masked, allowing the sweet, fruity smell of the ester to be easily detected.

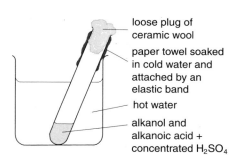

Figure 9.12 ▲
Making an ester.

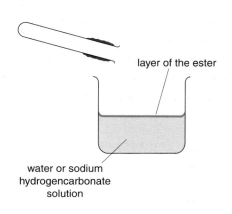

Figure 9.13 ▲
Separating the ester.

The simple esters are all sweet-smelling liquids and are widely used as fruit flavouring (e.g. pear drops) and in the cosmetics industry in perfumes, nail varnish and nail varnish remover. Industrially they are important solvents.

The formation of an ester from an alkanol and an alkanoic acid is an example of a **condensation reaction**. This is one in which two molecules join to form one larger molecule, with water formed at the same time. The hydroxyl functional group of the alkanol joins with the carboxyl functional group of the alkanoic acid to form the ester link in the ester molecule and water. We can see this in the equation for the reaction between propan-1-ol and ethanoic acid to give the ester propyl ethanoate and water.

ethanoic acid + propan-1-ol ⇌ propyl ethanoate + water

(ethanoate group) (propyl group)

Breaking esters – a hydrolysis reaction

The breaking down of an ester into an alkanol and an alkanoic acid by reaction with water is an example of a **hydrolysis reaction**. This is one in which a large molecule is broken down into smaller molecules by reaction with water.

In order to hydrolyse an ester, it is refluxed with sodium hydroxide solution rather than sulphuric acid. The reaction goes to completion using sodium hydroxide, but not with sulphuric acid because the reaction is reversible. The reason why the hydrolysis goes to completion using sodium hydroxide as catalyst is because sodium hydroxide is an alkali and reacts with the alkanoic acid as it is formed, preventing the reverse reaction from taking place.

The technique of refluxing (see Figure 9.14) is used for this reaction because simple esters are **volatile**, which means that they have low boiling points. They would therefore vaporise and easily escape from a beaker, for example, before the hydrolysis reaction could take place.

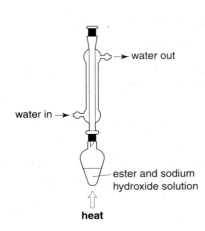

water out

water in →

ester and sodium hydroxide solution

heat

Figure 9.14 ▲
Apparatus for refluxing.

When the ester, methyl ethanoate, is hydrolysed using sodium hydroxide solution, the following reactions take place:

$$CH_3COOCH_3 + H_2O \rightarrow CH_3COOH + CH_3OH$$

methyl ethanoate   water   ethanoic acid   methanol

$$CH_3COOH \; + \; NaOH \; \rightarrow \; CH_3COONa \; + \; H_2O$$

ethanoic acid sodium hydroxide sodium ethanoate   water

Adding these two equations together gives, as an overall reaction:

$$CH_3COOCH_3 \; + \; NaOH \; \rightarrow \; CH_3COONa \; + \; CH_3OH$$

methyl ethanoate sodium hydroxide sodium ethanoate   methanol

After ester hydrolysis using an alkali such as sodium hydroxide solution, the alkanol, in this case methanol, can be distilled off and the alkanoic acid reformed by the addition of hydrochloric acid.

$$CH_3COONa + HCl \rightarrow CH_3COOH + NaCl$$

Making and breaking acids – A summary

condensation $\rightarrow$

(concentrated sulphuric acid catalyst)

alkanol + alkanoic acid $\rightleftharpoons$ ester + water

(sodium hydroxide solution catalyst)

$\leftarrow$ **hydrolysis**

Questions

9 (a) The reaction shown above is said to be 'reversible'. State what this means.
 (b) Name the ester that is formed by reaction between methanol and pentanoic acid.
 (c) What catalyst is used during ester formation reactions between an alkanol and an alkanoic acid?

10 (a) What type of reaction takes place when an ester is broken down by reaction with water to form an alkanol and an alkanoic acid?
 (b) Name the alkanol and alkanoic acid which can be formed by the breakdown of ethyl butanoate.
 (c) What catalyst ensures the complete breakdown of an ester into an alkanol and the sodium salt of an alkanoic acid?

Chapter 9 Study Questions

In questions 1–4 choose the correct word(s) **from the following list** to complete the sentence.

alkanes, alkenes, cycloalkanes, oxygen, carbon dioxide, addition, fermentation, condensation, hydrolysis.

1 A carbon-to-carbon double covalent bond is present in _____.

2 Esters can be made from alkanols and alkanoic acids by a _____ reaction.

3 When glucose breaks down to give ethanol in the presence of enzymes from yeast, the gas given off is _____.

4 Ethanol for industrial use is now made by catalytic hydration of ethene, but all ethanol for alcoholic drinks is made by _____.

5 The compound shown below

$$CH_3-CH_2-C-O-CH_2-CH_3$$
$$\underset{O}{\overset{\|}{}}$$

is formed by reaction between
A ethanol and butanoic acid
B propan-1-ol and ethanoic acid
C butan-1-ol and ethanoic acid
D ethanol and propanoic acid

Questions 6, 7 and 8 refer to the following reaction types:
A Hydration
B Addition
C Cracking
D Condensation

6 $C_6H_{12} + H_2 \rightarrow C_6H_{14}$

7 $C_6H_5COOH + CH_3OH \rightarrow C_6H_5COOCH_3 + H_2O$

8 $C_{14}H_{30} \rightarrow C_8H_{18} + C_6H_{12}$

9 Which of the following does **not** involve a chemical reaction?
A Hydrolysis
B Fermentation
C Distillation
D Combustion

10 The reaction between ethene and water to produce ethanol is important industrially, but does not take place readily.

(a) Write a chemical equation for this reaction using molecular formulae.
(b) What could be done in order to help the reaction take place more quickly?

11 (a) Why is hexane described as 'saturated', whereas hex-1-ene is described as 'unsaturated'?
(b) The shortened structural formula for hex-1-ene is as follows:

$$CH_2=CHCH_2CH_2CH_2CH_3$$

Give the molecular formula for the product that is obtained when hex-1-ene reacts with bromine in the test for unsaturation.

12 The table gives the heat given out in kilojoules (kJ) when one mole of certain alkanols burns.

Alkanol	Heat given out per mole (kJ)
methanol	730
ethanol	1370
propan-1-ol	2020
butan-1-ol	2670

(a) Draw a bar graph to show this information.
(b) What conclusion can be reached about the relationship between the size of the alkanol molecules and the energy given out per mole when they burn?

13 Ethanol may be dehydrated to give ethene. Give a shortened structural formula equation for this reaction.

14 The following compound can be made by reacting citric acid with methanol.

Circle and label **one** example of each of the following in the molecule:

(a) A hydroxyl group
(b) An ester link
(c) A carboxyl group

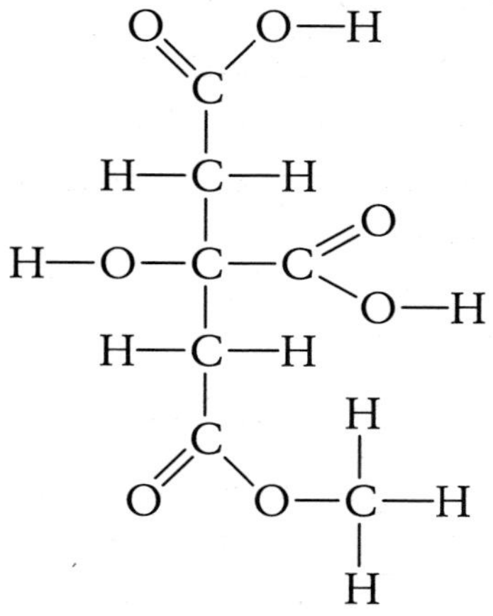

15 During cracking the finely powdered catalyst becomes covered in carbon and must be regenerated.

(a) Name a catalyst used during cracking.
(b) What effect does the carbon have on the efficiency of the catalyst?
(c) Suggest how the carbon can be removed from the catalyst so that it can be regenerated.

10 Plastics and Synthetic Fibres

Figure 10.1 ▲
Many parts of this Rollerblade are made of plastics.

Figure 10.2 ▲
Helmets are made of Kevlar.

Crude oil is the source of many useful materials, some of which are made into plastics and synthetic fibres. It is difficult to imagine our modern lives without either of these. A huge range of products contain plastic materials, from Rollerblades to mobile phones. Bottles, containers and packaging are often made of plastic. There are many different plastics, all with different properties.

If a bottle needs to be flexible, then polythene is used. If strength and transparency are important, then polyester is the plastic chosen. Where lightweight protection is required for packaging, polystyrene is ideal. Other widely used plastics include perspex, polyvinyl chloride (PVC), polypropene, Teflon, polyesters, nylons and silicones.

All of the plastics referred to above have been made by chemists since the beginning of the twentieth century. They are often preferred to traditional materials, such as wood, metal or paper, because of the better properties they have. Research into even more sophisticated plastics continues and has resulted in the production of the extremely strong and fire-resistant Kevlar and Nomex. There are now water-soluble plastics such as poly(ethenol) and plastics which conduct electricity.

A synthetic fibre is one which is not a natural product but is man-made. They are made by melting a plastic, or dissolving one in a suitable solvent, and then forcing the liquid through small holes. Labels on clothing indicate the presence of synthetic fibres with words such as polyester or polyamide. The word 'polyamide' indicates that nylon fibres are present. Very strong Kevlar fibres are obtained by dissolving Kevlar in concentrated sulphuric acid and then forcing the solution through small holes into water.

Synthetic and natural materials compared

For some uses, synthetic materials have advantages over natural materials, and vice versa. For example, polyester sportswear for athletes allows water to evaporate from the skin, whereas cotton absorbs moisture becoming damp and uncomfortable. For normal leisurewear, however, a cotton T-shirt is both cool and soft. A polypropene carpet is very hard-wearing and stain-resistant, whereas a wool carpet is not as tough and stains easily. A wool carpet may, however, be preferred because its fibres accept a dye better and it is both warmer and softer to the touch.

Figure 10.3 ▲
Polyester sportswear helps keep Henrick Larsson as cool as possible!

Figure 10.4 ▲
This bottle is both lightweight and shatterproof.

The same can be said when comparing jumpers made of synthetic fibres, such as Orlon, with those made of wool. Most people prefer the jumper made of wool.

Plastics find so many applications because they have useful properties. Most plastics are low density, are good heat and electrical insulators and are water-resistant. Individual plastics may have special properties that make them particularly suitable for certain applications. PVC, for example, is used in place of wood for window frames because it is tough, does not rot and does not require regular painting. The same plastic is used in place of iron for gutter pipes around houses because it does not rust and, like PVC window frames, it does not require painting. Plastic bottles made of polythene or polyester, for example, are preferred to ones made of glass because they are lighter and are shatterproof. In the case of the polythene bottles used for washing up liquid, the property of 'squeeziness' is also important.

Fish and chips were traditionally wrapped in newspaper, presumably because of its heat insulating properties, but now most fast food retailers use containers made of the even better heat insulator, polystyrene. Kevlar has replaced wood in hockey sticks and other materials in tennis racquets because it is lightweight, strong and flexible. Polythene pipes are now used to carry mains water (coloured blue) and mains gas (coloured yellow). Unlike the iron pipes they have replaced, they do not rust.

Questions

1 What is the meaning of the word 'synthetic'?

2 Draw a table to classify the following as being either 'natural' or 'synthetic': wool, nylon, polyester, polypropene, cotton.

3 Give three reasons, based on differences in properties, to explain why PVC tends to be used in place of wood for window frames.

Plastics and pollution

Most plastics are durable, which means that they last for a very long time. This is often a useful property, but it can also cause problems. Plastic litter can remain in the environment for a long time, whereas most natural materials, like wood, paper and cardboard, are **biodegradable** – they rot or decay away in time. Since most plastics do not rot away, they are said to be **non-biodegradable**. The low density of plastics can be an added problem. Crisp packets and other plastic waste, such as polythene sheet, can be blown about and left caught in trees and bushes.

Figure 10.5 ▲
Plastic waste causes environmental problems.

As a solution to the problem of plastics in the environment chemists have developed some which are biodegradable. One such biodegradable plastic is Biopol, which is not durable, but decays and reduces the litter problem.

It is a waste of resources to use a polythene carrier bag only once. It can be reused many times and then it should be recycled, rather than discarded. Polythene waste can be recycled as bin-liners, for example. Polystyrene waste can be recycled as flowerpots. There are, of course, many types of plastic and it is therefore important to sort them before recycling. Polythene bottles, for example, should be separated from polyester ('PET') bottles.

When anything burns some pollution is caused, such as smoke, for example. The burning of plastics and synthetic fibres, however, can often cause more pollution than the burning of natural materials such as wood, paper or cotton. All plastics and synthetic fibres contain carbon and hydrogen in their molecules and, when combustion is complete, produce carbon dioxide and water. However, lack of oxygen can lead to incomplete combustion with poisonous carbon monoxide being formed. Elements, other than carbon and hydrogen, which are present in some plastics and synthetic fibres can lead to the production of other toxic gases when they burn or smoulder. For example, polyurethane foam, which is used as padding in chairs and settees, contains nitrogen and forms toxic hydrogen cyanide (HCN) gas when it burns. Polyvinyl chloride (PVC) contains chlorine and produces the toxic, acidic gas hydrogen chloride (HCl) on combustion (see Table 10.1).

Elements present	Gas produced	Example of plastic
C	carbon monoxide, CO	Polystyrene
H and Cl	hydrogen chloride, HCl	Polyvinyl chloride (PVC)
H, C and N	hydrogen cyanide, HCN	Polyurethane

Table 10.1 ▲

Questions

4   Some crisp packets are made of polypropene lined with aluminium foil They are known to be non-biodegradable and difficult to recycle.

 (a) What is meant by the term 'non-biodegradable'?
 (b) Why is it important to recycle plastics if possible?

5 (a) All plastics can produce a poisonous gas when they burn incompletely because they contain carbon. Name this gas.
 (b) Polyurethane foam contains the elements carbon, hydrogen and nitrogen. What poisonous gas can be produced on combustion due to the presence of these elements?

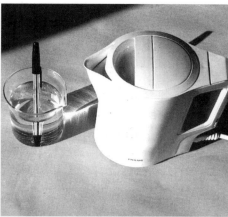

Figure 10.6 ▲
Ballpoint pen cases are made of polystyrene, which is a thermoplastic.

Thermoplastics and thermosetting plastics

Some plastics soften and melt when heated while others do not. A plastic which behaves in this way can easily be recycled as it can be moulded into a new shape, but it may not be suitable for use where high temperatures are found.

Thermoplastics soften on heating and are usually quite pliable. **Thermosetting plastics** do not soften on heating and tend to be hard and rigid.

Examples of thermoplastics include polythene, polystyrene, perspex, PVC and nylon. All thermoplastics will withstand moderate heating before they soften. Perspex, for example is widely used for making baths and wash basins, so it must be able to withstand hot water without softening. With care you can show that the polystyrene case of a ballpoint pen will soften and can be remoulded into a new shape if placed in boiling water. A simple test to distinguish between a thermoplastic and a thermosetting plastic is to touch it with a hot iron nail. If the plastic softens and melts, then it is a thermoplastic.

Figure 10.7 ▲
The plug and socket are made of thermosetting urea-methanal, but the wires are covered with thermoplastic PVC. What would happen if the wires became hot?

Thermosetting plastics cannot be reshaped once they have been moulded into a particular form and are therefore more difficult to recycle. Examples include formica and melamine which are used for kitchen worktop surfaces. Dark brown Bakelite was one of the first thermosetting plastics to be made and was widely used for electrical fittings. Urea-methanal has largely taken the place of Bakelite because its normally white appearance is preferable to the dark brown of Bakelite and it can be dyed to give any desired colour. Plug sockets are usually white urea-methanal, but the plugs that go into them, although made of the same thermosetting plastic, can be any colour.

Making plastics

The molecules which make up plastics are extremely large and are called **polymers** (poly = many, mers = parts). They are made by joining together many much smaller molecules called **monomers** (mono = one). The process of making a giant polymer molecule by joining together many small monomer molecules is called **polymerisation**.

There are two types of polymerisation, **addition polymerisation** and **condensation polymerisation**. During addition polymerisation the monomer molecules join to form polymer molecules and nothing else. When condensation polymerisation takes place, the monomer molecules join to form polymer molecules with water, or some other small molecule, formed at the same time.

The monomer molecules that are needed for both types of polymerisation are made from chemicals that come from crude oil.

Addition polymerisation

The monomers that undergo addition polymerisation are all unsaturated – they contain a carbon-to-carbon double covalent bond, C=C. When crude oil fractions are cracked, alkenes like ethene are produced and these, or derivatives of them such as chloroethene, are used to make addition polymers.

The modern name for an addition polymer is obtained by putting 'poly' in front of the bracketed monomer name. For many polymers, however, the old names are still in common use. For example, poly(ethene) is the name for the addition polymer made from the monomer ethene, but it is also known by its former names of polyethylene and polythene.

Figure 10.8 ▲
Addition polymerisation of ethene to give poly(ethene) takes place in this plant at Grangemouth on the Firth of Forth.

Monomer name	Polymer name
Ethene	poly(ethene)
Propene	poly(propene)
Chloroethene	poly(chloroethene)
Tetrafluoroethene	poly(tetrafluoroethene)
Phenylethene	poly(phenylethene)

Figure 10.9 ▲
Shopping bags are a common use for poly(ethene), which is also called polythene.

Poly(ethene)

Poly(ethene) was first made in England in the 1930s and became a secret military material during the Second World War when it was used as an electrical insulator for high frequency cables required for RADAR. From an annual production of a few hundred tonnes at that time it has become the most produced of all plastics with an annual world production of more than 15 million tonnes. There are several different types of poly(ethene) but the basic polymerisation reaction is common to all addition polymers.

During the reaction one of the bonded pairs of electrons in the C=C bond breaks to give a free electron on each carbon atom. These can then combine with free electrons in other molecules and a chain of ethene molecules can be joined together.

$$\cdots \underset{H}{\overset{H}{C}}=\underset{H}{\overset{H}{C}} + \underset{H}{\overset{H}{C}}=\underset{H}{\overset{H}{C}} + \underset{H}{\overset{H}{C}}=\underset{H}{\overset{H}{C}} + \cdots$$

ethene monomers

$$\cdots \; {}^{\bullet}\underset{H}{\overset{H}{C}}-\underset{H}{\overset{H}{C}}{}^{\bullet} + {}^{\bullet}\underset{H}{\overset{H}{C}}-\underset{H}{\overset{H}{C}}{}^{\bullet} + {}^{\bullet}\underset{H}{\overset{H}{C}}-\underset{H}{\overset{H}{C}}{}^{\bullet} + \cdots$$

$$\cdots -\underset{H}{\overset{H}{C}}-\underset{H}{\overset{H}{C}}-\underset{H}{\overset{H}{C}}-\underset{H}{\overset{H}{C}}-\underset{H}{\overset{H}{C}}-\underset{H}{\overset{H}{C}}- \cdots$$

poly(ethene) polymer

The equation for the polymerisation of ethene may be simplified as follows, where 'n' is a large number (more than a thousand):

$$n \; \underset{H}{\overset{H}{C}}=\underset{H}{\overset{H}{C}} \longrightarrow \left(\underset{H}{\overset{H}{C}}-\underset{H}{\overset{H}{C}} \right)_n$$

Figure 10.10 ▲
A strong form of polythene is used in the protective gear worn by American football players.

Polythene has a wide variety of uses which include film and sheet packaging, electrical insulation, carrier bags, food containers, bottles, basins, bowls, buckets, sacks and pipes.

Figure 10.12 ▲
Poly(propene) is used both in synthetic grass for hockey pitches and in car bumpers.

Poly(propene)

Poly(propene), like poly(ethene), is made by the addition polymerisation of a simple alkene. It is a stronger and tougher plastic than poly(ethene) and softens at a higher temperature.

Poly(propene) is made by the addition polymerisation of propene monomers.

propene monomers

poly(propene) polymer

The equation may also be shown as:

Figure 10.12 ▲
The packaging for this printer is made of polystyrene – now called poly(phenylethene).

Poly(propene) is a strong, tough plastic with great versatility. As a thin sheet, it can be used as packaging – packets of biscuits are often wrapped in it. It can be made into fibres for string, clothing, carpets and synthetic grass for sports pitches. When blow-moulded it can be formed into a range of products from washing-up bowls to car bumpers. It can also be extruded into pipes.

Poly(propene) is often referred to by its older name, polypropylene.

Poly(phenylethene)

Poly(phenylethene) is a very common plastic better known by its original name, polystyrene. It is found in two quite different forms. One is a rigid plastic, but quite brittle, used for cups, cartons, packaging, ballpoint pen barrels, etc. The other form is called 'expanded polystyrene' and is produced when tiny polystyrene granules containing a little pentane are heated. The liquid pentane boils inside each granule, forcing its way out and making the granules expand. The expanded form is widely used for packaging and to provide insulation, for example in caravans and refrigerators.

$$\cdots \quad \underset{H}{\overset{H}{C}}{=}\underset{C_6H_5}{\overset{H}{C}} \quad + \quad \underset{H}{\overset{H}{C}}{=}\underset{C_6H_5}{\overset{H}{C}} \quad + \quad \underset{H}{\overset{H}{C}}{=}\underset{C_6H_5}{\overset{H}{C}} \quad + \quad \cdots$$

phenyl(ethene) monomers

$$\cdots \quad {-}\underset{H}{\overset{H}{C}}{-}\underset{C_6H_5}{\overset{H}{C}}{-}\underset{H}{\overset{H}{C}}{-}\underset{C_6H_5}{\overset{H}{C}}{-}\underset{H}{\overset{H}{C}}{-}\underset{C_6H_5}{\overset{H}{C}}{-} \quad \cdots$$

poly(phenylethene) polymer

or

$$n \quad \underset{H}{\overset{H}{C}}{=}\underset{C_6H_5}{\overset{H}{C}} \quad \longrightarrow \quad {\left(\!\!{-}\underset{H}{\overset{H}{C}}{-}\underset{C_6H_5}{\overset{H}{C}}{-}\!\!\right)}_{n}$$

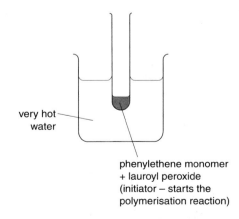

very hot water

phenylethene monomer + lauroyl peroxide (initiator – starts the polymerisation reaction)

Figure 10.13 ▲
This is how poly(phenylethene) can be made in the laboratory.

Poly(phenylethene) can be made in the laboratory by heating phenyl(ethene) and lauroyl peroxide in a test tube using a boiling water bath, as shown in Figure 10.13. Lauroyl peroxide is an initiator – it starts the polymerisation reaction. As polymerisation takes place the liquid monomer gradually thickens until it is a clear hard solid.

Poly(chloroethene)

Poly(chloroethene) is much better known by the older name of polyvinyl chloride or PVC. It is the most versatile plastic and, after poly(ethene), the most widely used. This is because substances called plasticisers can be added to it to give a large range of flexibility, from very rigid to very pliable.

chloro(ethene) monomers

poly(chloroethene) polymer

or

Figure 10.14 ▲
PVC is used in electrical cables, including this hair dryer

In its rigid form, poly(chloroethene), or PVC, is widely used for window frames. When made very flexible by the use of plasticisers it can be used for clothing, footwear, floor covering and wallpaper. The term 'vinyl' is often used to describe PVC when it is used for these flexible materials.

In common with most plastics PVC is a very good electrical insulator and the flexible form is used for this purpose on electrical wires and cables. Steel wire fencing is often covered with flexible PVC because it keeps out water and stops the steel from rusting.

PTFE – poly(tetrafluoroethene)

PTFE is better known by the brand name of Teflon. It has very low friction and is widely used for non-stick surfaces, on frying pans for example. Clothing treated with Teflon is stain resistant. It is made by the polymerisation of a monomer in which all four hydrogen atoms of an ethene molecule have been replaced by fluorine atoms. This molecule is then called tetrafluoroethene.

Figure 10.15 ▲
The Millennium Dome canopy is made of Teflon (PTFE)-coated glass fibre.

tetrafluoroethene monomers

poly(tetrafluoroethene) polymer

or

The canopy for the Millennium Dome covers a massive eight hectares and is made of Teflon (PTFE)-coated glass fibre. The Teflon encapsulates the glass, protecting it from water damage and abrasion.

Questions

8 State, in one sentence, the meaning of both of the words 'polymer' and 'monomer'.

9 In what way do the products of addition and condensation polymerisation differ?

10 Which bond is present in all monomers that can take part in addition polymerisation?

11 Using full structural formulae, write a chemical equation to show how three molecules of the monomer shown below can join to form an addition polymer.

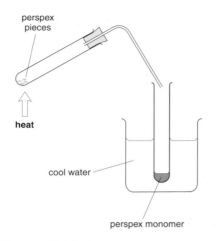

perspex
pieces

heat

cool water

perspex monomer

Figure 10.16 ▲
Perspex is easily depolymerised by heating.

Some other addition polymers

Many other addition polymers are in everyday use, apart from the ones already mentioned. They include Perspex, which is widely used as a lightweight replacement for glass in glazing. It can also be used for contact lenses and for spectacle lenses. As cast sheet it is used to make baths and washbasins.

When Perspex is heated strongly it readily depolymerises and turns back into its monomer. It is therefore very easy to recycle scrap Perspex.

Orlon and Courtelle are textile fibres made of the same addition polymer. Clothing made of this is hard wearing and washes easily. However, a jumper made of Orlon, for example, is generally considered to be less soft and warm than one made of the natural fibre wool.

Addition polymer 'backbones'

All addition polymer molecules have a 'backbone' made up entirely of carbon atoms.

$$...\ \text{—C—C—C—C—C—C—}\ ...$$

addition polymer 'backbone'

This means that addition polymer molecules can be readily distinguished from those of condensation polymers since the 'backbones' of the latter contain atoms of other elements as well as carbon.

Repeating units in addition polymers

The molecular structures of addition polymers made from a single type of monomer contain a repeated pattern of covalently bonded atoms called a **repeating unit**. The structural formula of the repeating unit is closely related to that of the monomer from which the polymer is formed. For the polymer poly(propene), for example, the structural formulae for the monomer and repeating unit are as follows:

propene monomer   repeating unit

The poly(propene) polymer molecule contains many such repeating units.

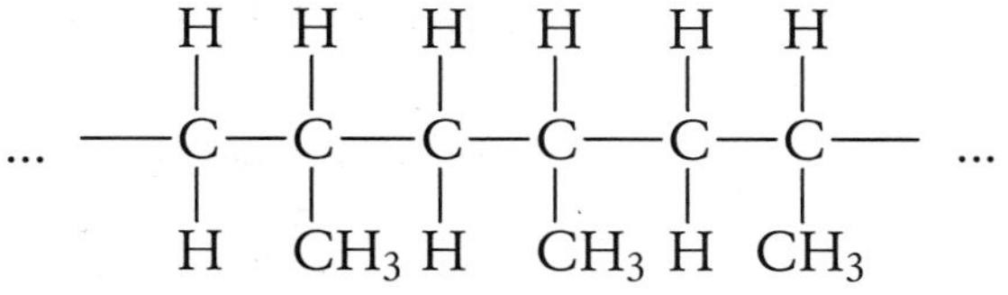

$$\ldots \; -\!\!\!\begin{array}{cccccc} H & H & H & H & H & H \\ | & | & | & | & | & | \\ C\!-\!C\!-\!C\!-\!C\!-\!C\!-\!C \\ | & | & | & | & | & | \\ H & CH_3 & H & CH_3 & H & CH_3 \end{array}\!\!\!- \; \ldots$$

repeating units

Addition polymers – A summary

Monomer		Polymer	Repeating unit
Ethene	$\begin{array}{c} H \quad\; H \\ \backslash \quad / \\ C\!=\!C \\ / \quad\; \backslash \\ H \quad\; H \end{array}$	Poly(ethene)	$\begin{array}{c} H \quad H \\ | \quad\; | \\ -C\!-\!C- \\ | \quad\; | \\ H \quad H \end{array}$
Propene	$\begin{array}{c} H \quad\; H \\ \backslash \quad / \\ C\!=\!C \\ / \quad\; \backslash \\ H \quad\; CH_3 \end{array}$	Poly(propene)	$\begin{array}{c} H \quad H \\ | \quad\; | \\ -C\!-\!C- \\ | \quad\; | \\ H \quad CH_3 \end{array}$
Phenylethene (Styrene)	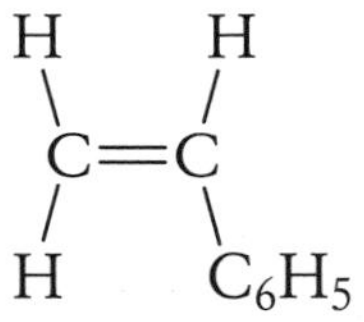 $\begin{array}{c} H \quad\; H \\ \backslash \quad / \\ C\!=\!C \\ / \quad\; \backslash \\ H \quad\; C_6H_5 \end{array}$	Poly(phenylethene) (Polystyrene)	$\begin{array}{c} H \quad H \\ | \quad\; | \\ -C\!-\!C- \\ | \quad\; | \\ H \quad C_6H_5 \end{array}$
Chloroethene (Vinyl chloride)	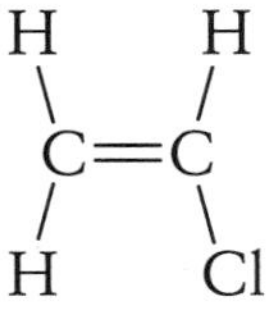 $\begin{array}{c} H \quad\; H \\ \backslash \quad / \\ C\!=\!C \\ / \quad\; \backslash \\ H \quad\; Cl \end{array}$	Poly(chloroethene) (Polyvinyl chloride)	$\begin{array}{c} H \quad H \\ | \quad\; | \\ -C\!-\!C- \\ | \quad\; | \\ H \quad Cl \end{array}$
Tetrafluoroethene	$\begin{array}{c} F \quad\; F \\ \backslash \quad / \\ C\!=\!C \\ / \quad\; \backslash \\ F \quad\; F \end{array}$	Poly(tetrafluoroethene)	$\begin{array}{c} F \quad F \\ | \quad\; | \\ -C\!-\!C- \\ | \quad\; | \\ F \quad F \end{array}$

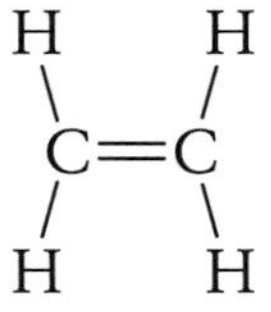

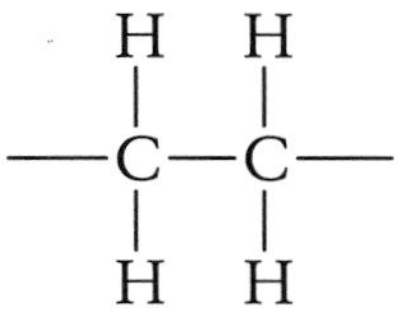

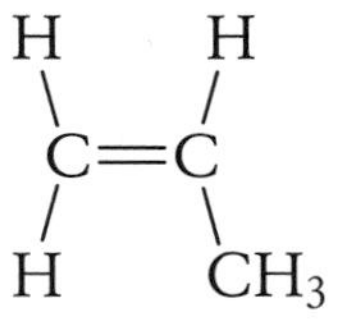

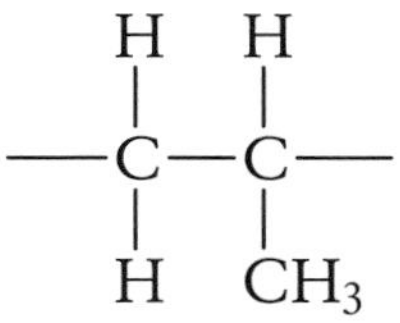

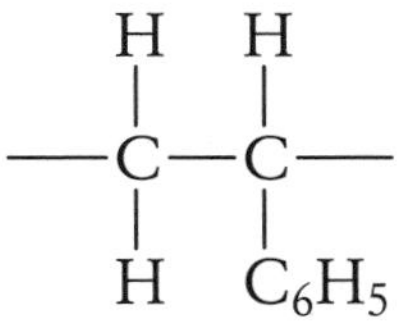

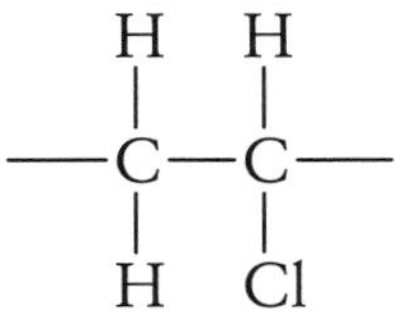

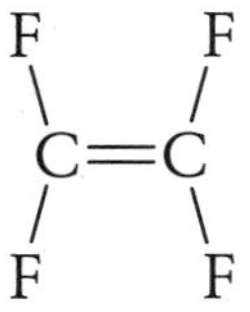

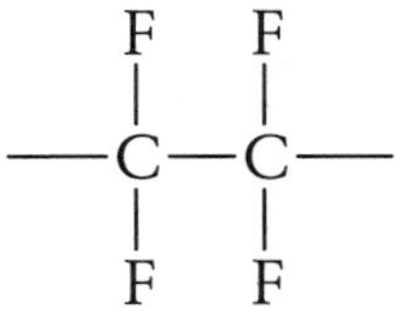

Table 10.2 ▲

Question

12 Perspex, a lightweight plastic replacement for glass in glazing, is made from the following monomer.

$$\begin{array}{ccc} H & & CH_3 \\ \diagdown & & \diagup \\ & C = C & \\ \diagup & & \diagdown \\ H & & COOCH_3 \end{array}$$

Draw **(a)** the repeating unit and **(b)** a part of the Perspex polymer chain showing three repeating units. Use full structural formulae for each.

The group of atoms shown as C_6H_4 in terephthalic acid is a ring structure which chemists show as a circle within a hexagon. The two carboxyl groups are attached at either side of the ring. Terephthalic acid molecules can therefore be drawn as follows:

Condensation polymerisation

When an addition polymer is made from its monomer molecules, all of the atoms in the monomers are used to make the polymer. Condensation polymerisation is different in that, when the monomer molecules join together to form the polymer, not all of the atoms in the monomers are used to make the polymer. When condensation polymerisation takes place, many monomer molecules join to form one large polymer molecule, but water, or some other small molecule, is formed at the same time.

The monomers that are used to make condensation polymers are different from those used to make an addition polymer. Usually only one type of monomer molecule is used to make an addition polymer and all such monomer molecules contain a carbon-to-carbon double covalent bond, C=C. Condensation polymers are usually made from two different monomer molecules and each of these must have two suitable functional groups, such as hydroxyl, –OH, carboxyl, –COOH, or amine, $-NH_2$.

Polyesters

The first commercially useful polyester to be manufactured was made from terephthalic acid, which contains two carboxyl groups, and ethane–1,2–diol, which contains two hydroxyl groups.

$$\begin{array}{cc} H{-}O{-}\underset{\underset{O}{\parallel}}{C}{-}C_6H_4{-}\underset{\underset{O}{\parallel}}{C}{-}O{-}H & H{-}O{-}CH_2{-}CH_2{-}O{-}H \end{array}$$

terephthalic acid   ethane–1, 2-diol

The condensation polymer which is formed by a reaction between terephthalic acid and ethane–1,2–diol is called poly(ethene terephthalate), PET, but it is much better known as Terylene.

Although many other polyesters have been made, Terylene is still the one which is most widely found. It is a very important textile fibre, which

terephthalic acid and ethane–1, 2-diol monomers

ester links

polyester polymer (Terylene)

repeating unit

can be produced as very fine microfibres used to make fashionable clothes and sportswear. The letters PET are to be found on many fizzy drinks bottles and Terylene is also used in videotapes, CDs, and DVDs.

As with addition polymers, the molecular structure and the repeating unit of a condensation polymer, such as a polyester, can be drawn if the structures of the monomers are provided. For a diol reacting with a dicarboxylic acid, the general reaction is as follows:

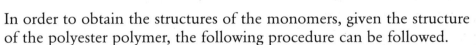

diol

carboxylic acid

polyester   polymer

repeating unit

Figure 10.17 ▲
Videotapes, CDs and DVDs all contain the polyester Terylene.

In order to obtain the structures of the monomers, given the structure of the polyester polymer, the following procedure can be followed.

(a) Locate the ester link:

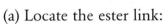

(b) Water adds on like this:

(c) The carboxyl and hydroxyl groups in the monomers can now be formed:

Polyamides

Polyamides are condensation polymers which are held together by the **amide link** $-CO-NH-$, the full structural formula of which is:

The best known polyamides are called **nylons**, which contain different numbers of $-CH_2-$ groups between the amide links. The number of these groups is indicated by numbers written after the word nylon. One of the most common nylons is nylon-6,6, which was also the first to be manufactured on a commercial scale. In common with many nylon polymers, it is made by a condensation reaction between a dicarboxylic acid, which contains two carboxyl groups $-COOH$, and a diamine, which contains two amine groups $-NH_2$. The monomers which are used to make nylon-6,6 are hexane-1,6-dioic acid and 1,6-diaminohexane.

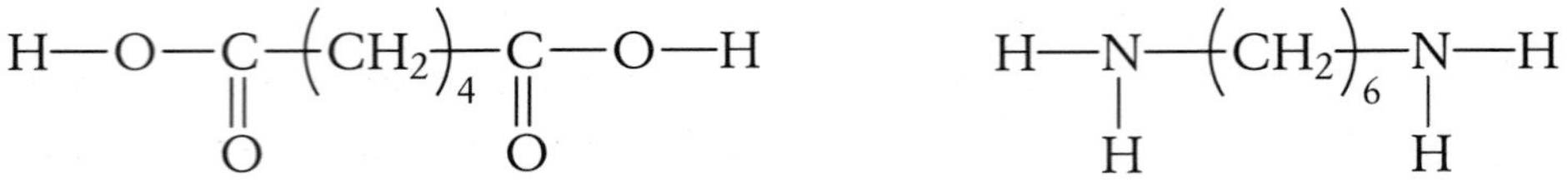

hexane-1,6-dioic acid   1,6-diaminohexane

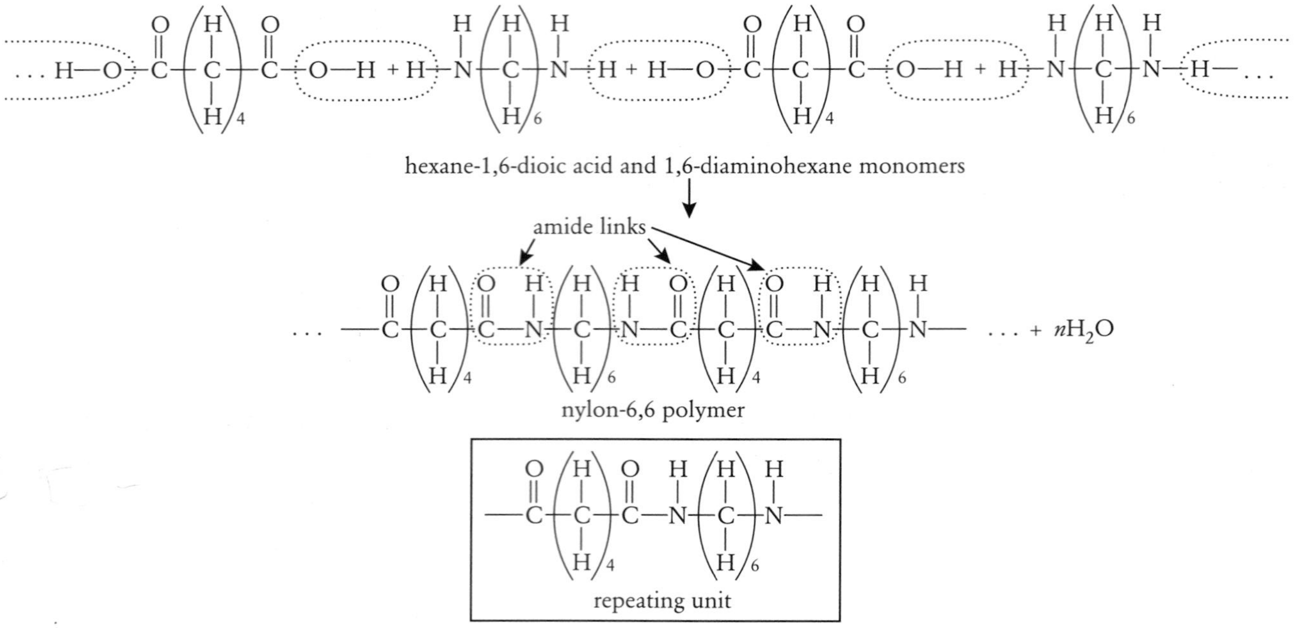

hexane-1,6-dioic acid and 1,6-diaminohexane monomers

amide links

nylon-6,6 polymer

repeating unit

Many polyamides are made by the condensation polymerisation of a dicarboxylic acid with diamine. The general reaction is as follows:

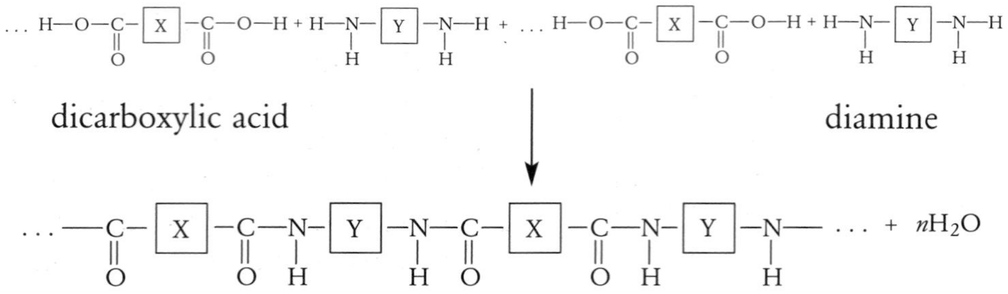

dicarboxylic acid   diamine

polyamide polymer

$$-C-\boxed{X}-C-N-\boxed{Y}-N-$$
$$\ \ \ \|\ \ \ \ \ \ \ \ \ \|\ \ \ |\ \ \ \ \ \ \ \ \ \ |$$
$$\ \ \ O\ \ \ \ \ \ \ \ \ O\ \ H\ \ \ \ \ \ \ \ \ \ H$$

repeating unit

When X and Y are both made up of CH_2 groups, the nylon polyamide polymers are thermoplastic and soften when heated. In the case of the polyamide known as Kevlar, however, both X and Y are made up of a bulky ring group, C_6H_4. Kevlar polymer chains are strongly attracted to each other with the result that it is a thermosetting polymer.

It is possible to obtain the structures of the monomers, given the structure of the polyamide polymer, using a similar method to that used with polyester polymers.

(a) Locate the amide link:

$$\ \ \ \ \ \ \ \ \ \ \ \ \ \ O\ \ H$$
$$\ \ \ \ \ \ \ \ \ \ \ \ \ \ \|\ \ \ |$$
$$\ldots\ -C-N-$$

(b) Water adds on like this:

$$\ \ \ \ \ \ \ \ \ \ \ \ \ \ O\ \ H$$
$$\ \ \ \ \ \ \ \ \ \ \ \ \ \ \|\ \ \ |$$
$$\ldots\ -C-N-\ \ldots$$
$$\ \ \ \ \ \ \ \ \ \ \ \ \ \ \Uparrow$$
$$\ \ \ \ \ \ \ \ \ \ \ \ O-H$$
$$\ \ \ \ \ \ \ \ \ \ \ \ \ \ |$$
$$\ \ \ \ \ \ \ \ \ \ \ \ \ \ H$$

(c) The carboxyl and amine groups in the monomers can now be formed:

$$\ \ \ \ \ \ \ \ O\ \ \ \ \ \ \ \ \ \ \ \ \ \ \ \ \ \ H$$
$$\ \ \ \ \ \ \ \ \|\ \ \ \ \ \ \ \ \ \ \ \ \ \ \ \ \ \ \ |$$
$$\ldots\ -C-O-H + H-N-\ \ldots$$

It should be noted that, when breaking either an ester link, or an amide link, the OH group from water adds on to the CO group producing a carboxyl group, COOH.

Making nylon in the laboratory

Nylons can be made in the laboratory by using a diacid chloride instead of a dicarboxylic acid. Nylon-6,6, for example, can be made using hexane-1,6-dioyl chloride and 1,6-diaminohexane. A solution of the chloride in cyclohexane is carefully added to form a layer on top of a solution of the diamine in water. Nylon-6,6 forms at the interface and can be removed as a continuous fibre by grasping the nylon with tweezers and winding it onto a glass rod (see Figure 10.19). In this case the polymerisation results in the formation of hydrogen chloride instead of water.

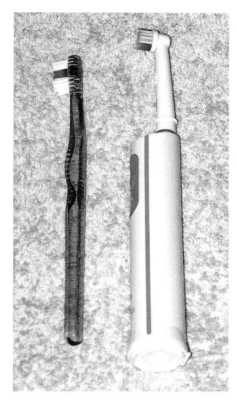

Figure 10.18 ▲
Toothbrush bristles are made of nylon.

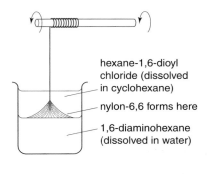

hexane-1,6-dioyl chloride (dissolved in cyclohexane)

nylon-6,6 forms here

1,6-diaminohexane (dissolved in water)

Figure 10.19 ▲
Laboratory preparation of nylon-6,6.

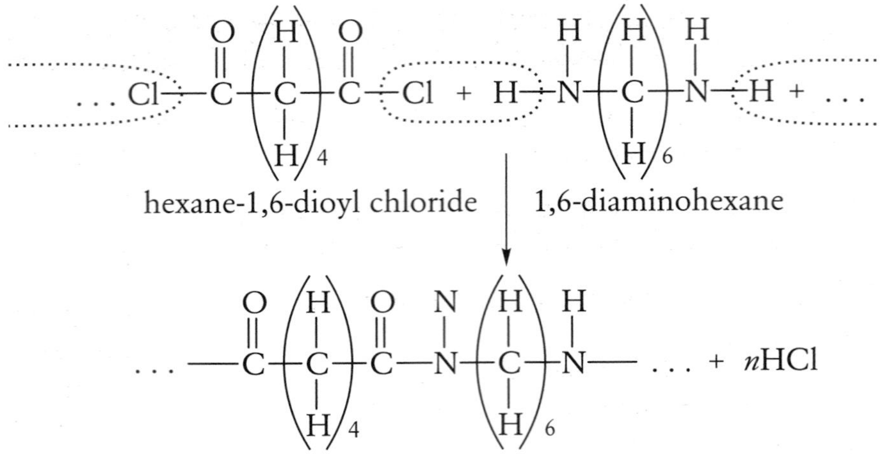

hexane-1,6-dioyl chloride | 1,6-diaminohexane

nylon-6,6

Condensation polymer 'backbones'

Unlike addition polymers that have a 'backbone' made up entirely of carbon atoms, condensation polymers have a backbone that contains other elements apart from carbon. Polyesters have oxygen atoms in their 'backbone' and polyamides have nitrogen atoms in it.

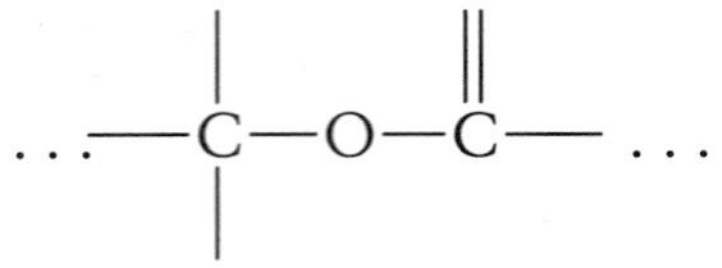

part of a polyester 'backbone' 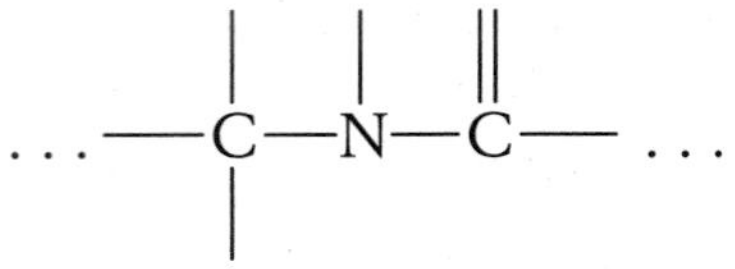   part of a polyamide 'backbone'

Questions

13 How many suitable functional groups *must* be present in a monomer molecule if it is to take part in condensation polymerisation?

14 Name and give full structural formulae for the functional groups present in the following monomers which are used to make a polyester: CH_2CHCH_2OH $HOOC$—◯—$COOH$.

15 Nylon-4,6 is made using the following monomer molecules.

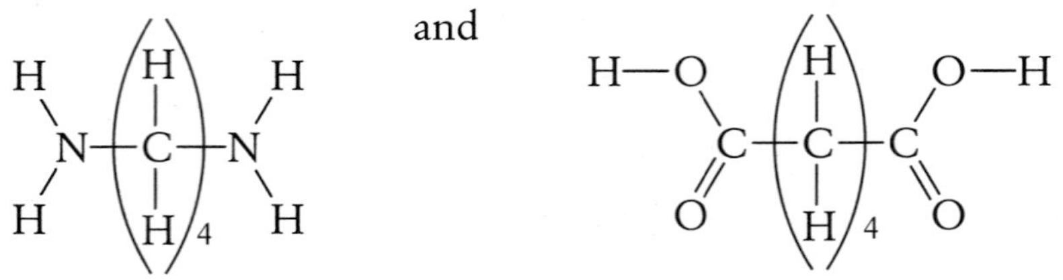

and

Draw a part of the structure of the nylon-4,6 polymer, showing two repeating units and circling an amide link.

Chapter 10 Study Questions

In questions 1–4 choose the correct word(s) **from the following list** to complete the sentence.

addition, condensation, thermoplastic, thermosetting plastic, saturated, unsaturated, amide link, ester link

1 Monomers that undergo addition polymerisation are _____.
2 All forms of nylon contain the _____.
3 If a plastic does not soften, even on strong heating, it must be a _____.
4 Polyesters are formed by _____ polymerisation.

5 From which of the following are most plastics and synthetic fibres made?
   A Coal   B Wood   C Oil   D Plants

6 Body armour needs to be lightweight but very strong. Which of the following plastics is most suitable for this use?
   A Polyester   B Perspex
   C Polyethenol   D Kevlar

7 Propene has the following structural formula.

$$\underset{H}{\overset{H}{\underset{|}{}}}C=C\underset{CH_3}{\overset{H}{}}$$

Which of the following is a part of the structure of poly(propene)?

A
$$-C-C-CH_2-C-C-CH_2-$$

B
$$=C-C=C-C=C-C=$$

C
$$-C-C-C-C-$$

D
$$-C-C-C-C-$$

8 An example of a degradable plastic is
   A Biopol
   B Nylon
   C Formica
   D Polystyrene

9 (a) Draw the full structural formula of ethene.
   (b) Draw a part of the poly(ethene) polymer molecule showing three monomer molecules joined together.
   (c) What type of polymerisation takes place when poly(ethene) is formed from ethene monomers?

10 The densities of some plastics are given in the table.

Plastic	Density (g cm^{-3})
Perspex	1.20
Polystyrene	1.05
Poly(propene)	0.90
Nylon–6,6	1.10

   (a) Draw a bar graph to show this information.
   (b) Which of the plastics referred to would float in water (density 1.00 g cm^{-3})?

11 Part of the structure of a form of nylon is shown below.

   (a) Draw the structure of the amide link.
   (b) Draw the repeating unit in this form of nylon.
   (c) Draw full structural formulae for the monomers from which this polymer is made.

12 Part of the structure of a polyester is shown below.

   (a) Draw the structure of the ester link.
   (b) Draw structural formulae for the monomers from which this polymer is made.

11 Natural Products

Carbohydrates

Figure 11.1 ▲
These foods are rich in carbohydrates.

Making carbohydrates

Carbohydrates are an important class of food made by plants during a process called photosynthesis. The plants take in carbon dioxide and water for this process, which involves a series of chemical reactions. The end products are carbohydrates, which stay in the plant, and oxygen that is released into the air. The plants store carbohydrates in tubers and seeds as an energy reserve. Foods with a high carbohydrate content therefore include potatoes and other root vegetables. Foods that are seeds, or are made from seeds, also have a lot of carbohydrate in them. They include rice, wheat and oats, which are used to make breakfast cereals. These are also used to make foods such as bread, cakes, biscuits and pasta. Pea and bean seeds are also rich in carbohydrates.

The simple carbohydrate, glucose, is made by plants during photosynthesis. The overall reaction can be shown as:

carbon dioxide + water → glucose + oxygen

The balanced formula equation is:

$$6CO_2 + 6H_2O \rightarrow C_6H_{12}O_6 + 6O_2$$

Like all carbohydrates, glucose contains the elements carbon, hydrogen and oxygen, with two hydrogen atoms for every oxygen atom, as in water.

a)

b)

c)

Figure 11.2 ▲
Concentrated sulphuric acid removes water from carbohydrates to leave just carbon.

Dehydrating carbohydrates

Adding concentrated sulphuric acid to a carbohydrate removes the hydrogen and oxygen atoms to leave a black mass of carbon (see Figure 11.2). This reaction takes place because concentrated sulphuric acid is a powerful dehydrating agent (remover of water). The balanced formula equation for the dehydration of glucose is as follows:

$$C_6H_{12}O_6 \rightarrow 6C + 6H_2O$$

absorbed by concentrated sulphuric acid

Using carbohydrates

Carbohydrates supply the body with energy (see Figure 11.3), the simpler ones combining with oxygen in exothermic reactions, called respiration, to produce carbon dioxide and water.

The overall reaction involving glucose can be shown as:

$$\text{glucose} + \text{oxygen} \rightarrow \text{carbon dioxide} + \text{water}$$

The balanced formula equation is:

$$C_6H_{12}O_6 + 6O_2 \rightarrow 6CO_2 + 6H_2O$$

Various experiments can be carried out to show that the reaction between a carbohydrate and oxygen is exothermic. In one experiment, shown in Figure 11.4, a dry carbohydrate-rich food, such as cornflakes, can be made to burn brilliantly in oxygen released by heating potassium permanganate crystals.

Figure 11.3 ▲
Carbohydrates provide athletes with the energy they need to run (and to jump).

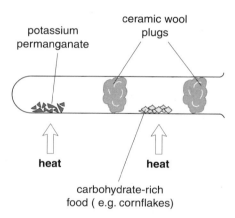

Figure 11.4 ▲
Carbohydrates burn well when heated in air or oxygen.

Questions

1 Why are carbohydrates important in our diet?

2 Which of the following could be the molecular formula of a carbohydrate? C_2H_6O; $C_{18}H_{36}O$; $C_{12}H_{22}O_{11}$; $C_4H_8O_5$.

3 What word describes a reaction in which heat energy is given out?

4 Describe the tests for carbon dioxide and water.

Carbohydrates large and small

The carbohydrates that we use as foods can be divided into two groups. The first group consists of the sweet-tasting, water-soluble carbohydrates that we call **sugars**. These are made of fairly small molecules. Glucose and fructose are common simple sugars with the same molecular formula $C_6H_{12}O_6$. Sucrose (table sugar) and maltose are made of molecules which are almost twice as large since they both have molecular formula $C_{12}H_{22}O_{11}$ (see Table 11.1).

Sugar	Where found	Molecular formula
glucose	grapes, honey	$\left.\right\} C_6H_{12}O_6$
fructose	many fruits, honey	
sucrose	sugar cane, sugar beet	$\left.\right\} C_{12}H_{22}O_{11}$
maltose	sprouting barley	

Table 11.1 ▲
Some sugars and where they are found.

Figure 11.5 ▲
Honey contains equal masses of glucose and fructose.

The other group of carbohydrates that we use as food consists of one single member called **starch**. Starch is a tasteless, powdery compound, which does not truly dissolve in water, although it can be dispersed to give a cloudy mixture, which is then called 'starch solution'.

Starch is a **natural condensation polymer** made by plants from glucose and used as a store for energy.

Using H–$\boxed{G}$–O–H to represent glucose, the polymerisation is:

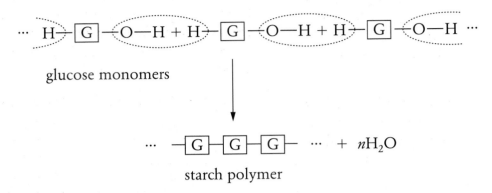

glucose monomers

starch polymer

Using chemical formulae, the balanced equation for the polymerisation of glucose to give starch is:

$$nC_6H_{12}O_6 \rightarrow (C_6H_{10}O_5)_n + nH_2O$$

glucose   starch

In this equation n is a number which can have a value in the range of about 60 up to about 600.

Comparing sugars and starch

A beam of light passes straight through a sugar solution with no trace of its path left behind. A light beam is reflected back by 'starch solution' so that the path of the beam can be seen. This can be shown by a simple experiment (see Figure 11.6).

The reason the beam of light can be seen in the 'starch solution' is that the starch molecules are big enough to reflect light. The much smaller sugar molecules cannot do this.

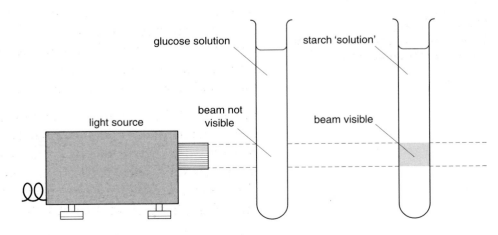

Figure 11.6 ▲
Starch molecules are large enough to reflect light, sugar molecules are not.

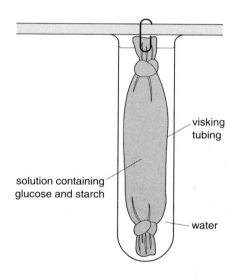

Figure 11.7 ▲
Glucose molecules can pass through the visking tubing but starch cannot.

Classifying the carbohydrates

Chemists classify carbohydrates according to the size of their molecules. Those with molecular formula $C_6H_{12}O_6$, such as glucose and fructose, are called **monosaccharides** Those with molecular formula $C_{12}H_{22}O_{11}$, such as sucrose and maltose, are called **disaccharides**. Starch is referred to as a **polysaccharide** because it has molecular formula $(C_6H_{10}O_5)_n$, where n is a large number. The full classification is given in Table 11.2.

Type of carbohydrate	Molecular formula	Example(s)
monosaccharide	$C_6H_{12}O_6$	glucose, fructose
disaccharide	$C_{12}H_{22}O_{11}$	sucrose, maltose
polysaccharide	$(C_6H_{10}O_5)_n$	starch

Table 11.2 ▲
Classfication of carbohydrates.

Chemical tests for carbohydrates

Most sugars, including glucose, fructose and maltose, produce an orange-red colour when heated with **Benedict's solution**. The heating is most safely carried out in a water bath. As the reaction proceeds, the blue Benedict's solution goes through a variety of colours before the production of the orange-red colour. Sucrose does *not* give this result and is unaffected by Benedict's solution, which stays blue in colour.

Starch can be easily distinguished from other carbohydrates by the addition of **iodine solution** with which it produces a dark blue colour. The original colour of the iodine solution is reddish-brown. No other carbohydrate gives this result.

Breaking down carbohydrates

If you chew a piece of bread for about ten minutes, it will eventually taste of sugar. This is because bread contains starch and, as you chew, this is broken down by reaction with water (hydrolysed) to produce maltose. On complete digestion, starch is hydrolysed to glucose. This is important because glucose can pass through the gut wall and into the bloodstream. The much larger starch molecules cannot do this.

In Figure 11.7, a mixture of glucose solution and 'starch solution' are placed inside some visking tubing. This is made from a plastic that contains tiny pores, just like the gut wall. At regular intervals the water in the test tube is tested for starch and for glucose. It is found that glucose can pass through the visking tubing, but starch cannot. The same thing happens at the gut wall.

Glucose is carried by the bloodstream to body cells where it reacts with oxygen to produce carbon dioxide, water and, more importantly, energy.

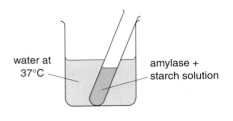

water at 37°C — amylase + starch solution

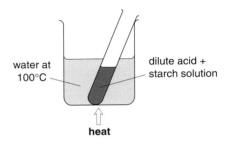

water at 100°C — dilute acid + starch solution

heat

Figure 11.8 ▲
Two ways of hydrolysing starch.

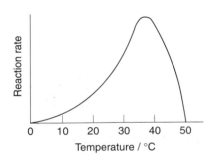

Figure 11.9 ▲
The effect of temperature on an enzyme-catalysed reaction.

Two ways of breaking down starch

Amylase is an enzyme found in saliva. It catalyses the hydrolysis of starch to maltose. Body enzymes, like amylase, work best at body temperature, which is about 37°C. As a result, if a mixture of amylase and 'starch solution' is kept at about 37°C then the starch is hydrolysed very quickly.

$$(C_6H_{10}O_5)_n + n/2H_2O \rightarrow n/2C_{12}H_{22}O_{11}$$

starch   maltose

Starch can also be broken down into simple sugar molecules by heating it in a dilute acid solution, such as dilute hydrochloric acid. In this case the starch is hydrolysed quite slowly to give glucose molecules, despite the higher temperature.

$$(C_6H_{10}O_5)_n + nH_2O \rightarrow nC_6H_{12}O_6$$

starch   glucose

The two experiments are shown in Figure 11.8. In order to follow the experiments, samples are removed from the test tubes at intervals and tested with Benedict's solution. In the case of the acid-catalysed reaction, the samples must be neutralised using an alkali before testing with Benedict's solution.

The temperature at which an enzyme works best is called its **optimum temperature**. For body enzymes the optimum temperature is the normal body temperature of about 37°C. At higher temperatures they are quickly destroyed, while at lower temperatures they are less efficient. The effect of temperature on a reaction catalysed by a body enzyme is shown in Figure 11.9.

Prescribed Practical Activity

Hydrolysis of starch

- The aim of this experiment is to hydrolyse starch in the presence of either an enzyme or an acid and to demonstrate that the enzyme or acid catalyses the reaction.
- The enzyme-catalysed reaction is the most straight-forward.
- Apparatus as shown in Figure 11.8 above can be used, but a test tube containing starch solution only should also be present as a control.
- After just a few minutes, the enzyme (amylase)

catalyses sufficient hydrolysis of the starch for a positive test with Benedict's solution to be obtained by raising the temperature of the water in the beaker.

- The control does not give a positive test with Benedict's solution, proving that it must be the enzyme that caused the hydrolysis reaction – it was acting as a catalyst.
- In this reaction, starch is hydrolysed to give a sugar which gives a positive test with Benedict's solution.

Figure 11.10 ▲
Some protein-rich foods.

Questions

8 Name three carbohydrates that give a positive test with Benedict's solution.

9 Name two carbohydrates that do not give a positive test with Benedict's solution.

10 Describe the initial and final colours during a positive test with Benedict's solution.

11 Name a carbohydrate that gives a positive test with iodine solution.

12 Describe the initial and final colours during a positive test with iodine solution.

13 Acids and enzymes can hydrolyse starch. Which are the more efficient?

14 How could you show that a sugar had been produced by the hydrolysis of starch? Give the test and the result.

Proteins

Proteins in our diet

Proteins are an important class of food made by plants. All proteins contain the elements carbon, hydrogen, oxygen and nitrogen. Plants take the nitrogen that they require to make proteins from the soil in the form of nitrates. Foods with a high protein content include fish, meat, cheese, eggs, peas, beans and lentils.

Using proteins

Proteins make up about 15% of the human body by mass. They are the major structural material of animal tissue. Muscle, tendons, skin, nails and hair are all mainly protein, with the very long thin protein molecules having a fibrous nature.

Proteins are also involved in the maintenance and regulation of life processes. These are globular proteins. For example, the protein haemoglobin is the red pigment and oxygen carrier in blood. This and some other, less obvious and more specialised functions of proteins are shown in Table 11.3.

Protein	Function
enzymes	act as catalysts
antibodies	confer immunity to disease
haemoglobin	transports oxygen
insulin (a hormone)	promotes glucose use

Table 11.3 ▲
Some specialised functions of proteins.

In 1954 **insulin** became the first protein to have its amino acid sequence worked out. This followed ten years of research by **Frederick Sanger**, who was later awarded the **Nobel Prize for Chemistry** (in 1958). He showed that insulin molecules are made from 51 amino acid molecules in total and that 17 *different* amino acids are present. He also worked out the precise sequence of all of the amino acid molecules in the insulin molecule.

The structure of protein molecules

Proteins are condensation polymers made of many **amino acid** molecules joined together. Insulin is one of the smaller proteins, having 51 amino acids joined together. In human haemoglobin there are 578 amino acids and in the enzyme, urease, there are 4500.

Amino acids contain two functional groups, namely the amine group $-NH_2$ and the carboxyl group –COOH. The amino acids used to make proteins all have the following molecular structure, where **R** stands for the rest of the molecule:

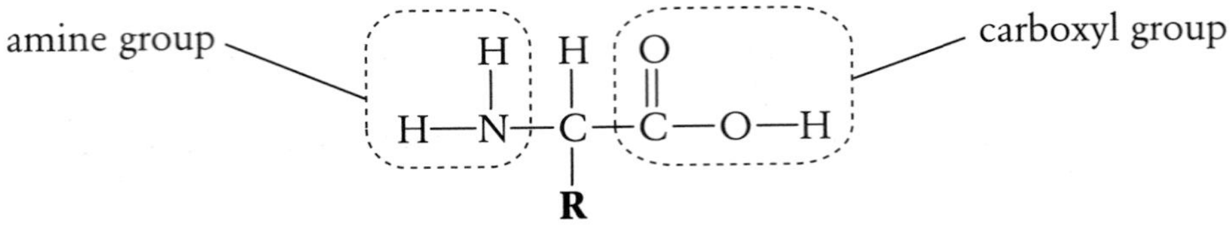

Only 20 amino acids are commonly found in proteins, while another six occur occasionally. The molecular structures of three of the simpler amino acids are shown in Table 11.4.

Name	R group	Full structural formula
glycine	—H	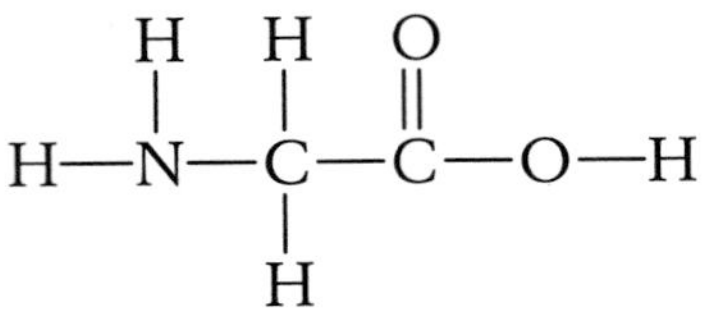
alanine	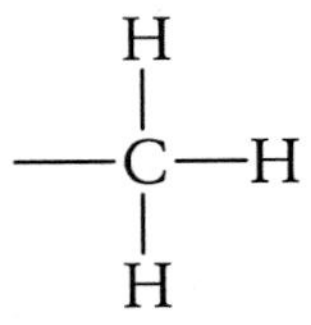	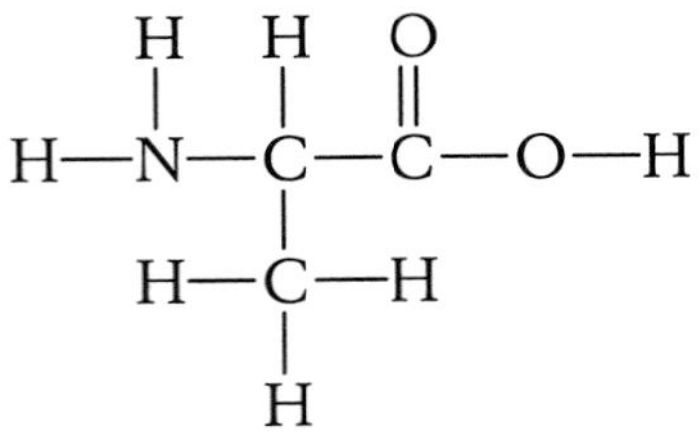
serine	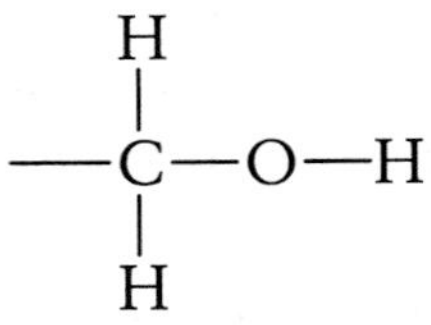	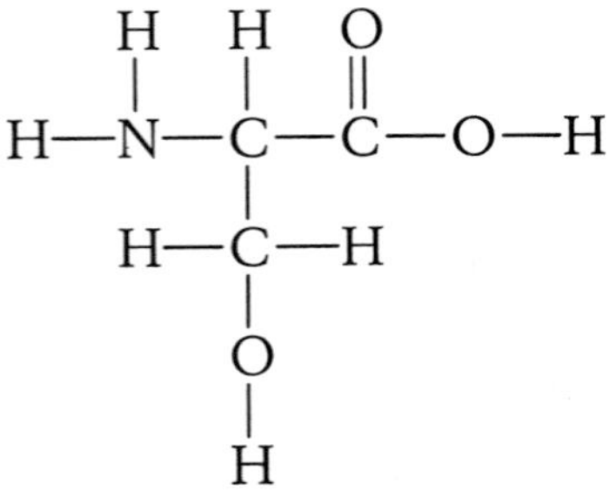

Table 11.4 ▲
Some simple amino acids

Condensation polymerisation of amino acids

As with polyamide formation, protein formation is due to a condensation reaction between amine groups ($-NH_2$) and carboxyl groups ($-COOH$). The condensation polymerisation of amino acids during protein formation produces the same link, $-CO-NH-$, that is present in polyamides. However, whereas in polyamides it is called the amide link, in proteins it is called the **peptide link**.

The joining of the three amino acids shown in Table 11.4, along with other amino acids, to form part of a protein molecule can be shown as follows.

peptide links

Proteins can contain several thousand amino acids joined together. There is a huge variety of proteins in plants and animals which is made possible by the fantastic variety of possible ways of arranging up to 26 different amino acids in varying numbers and varying sequences. You can imagine the amino acids as being like the letters of the alphabet and the great variety of proteins as being 'words' made up from the amino acid 'letters'.

Hydrolysis of proteins

During digestion, enzymes catalyse the hydrolysis of proteins in our diet to produce amino acids. Like glucose molecules produced by the hydrolysis of starch, amino acids can pass through the gut wall and into the bloodstream. The blood carries the amino acids to body cells where they are built up into proteins that are specific to the body's needs.

In the laboratory, proteins can be hydrolysed to produce amino acids by refluxing with hydrochloric acid (see Figure 11.11).

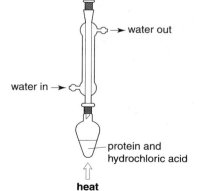

Figure 11.11 ▲
Apparatus for refluxing.

The structural formulae of amino acids formed by the hydrolysis of proteins can be identified from the structure of a section of the protein by the following procedure:

$$\cdots \overset{\overset{\displaystyle O}{\|}}{—C}—\overset{\overset{\displaystyle H}{|}}{N}— \cdots$$

Locate a peptide link:

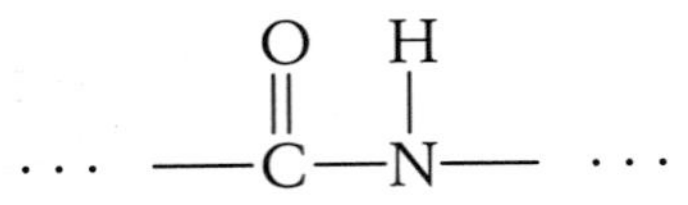

Add water like this:

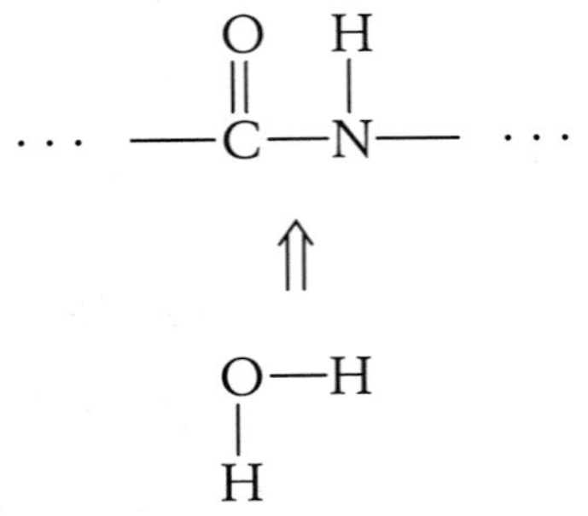

$$\cdots \overset{\overset{\displaystyle O}{\|}}{—C}—\overset{\overset{\displaystyle H}{|}}{N}— \cdots$$

$$\Uparrow$$

$$\overset{\displaystyle O—H}{\underset{\displaystyle H}{|}}$$

The carboxyl and amine groups can now be reformed:

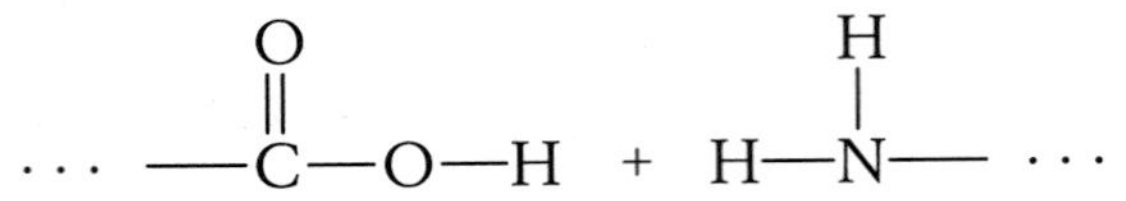

$$\cdots \overset{\overset{\displaystyle O}{\|}}{—C}—O—H \quad + \quad H—\overset{\overset{\displaystyle H}{|}}{N}— \cdots$$

An equation for the hydrolysis of part of a protein chain is shown below.

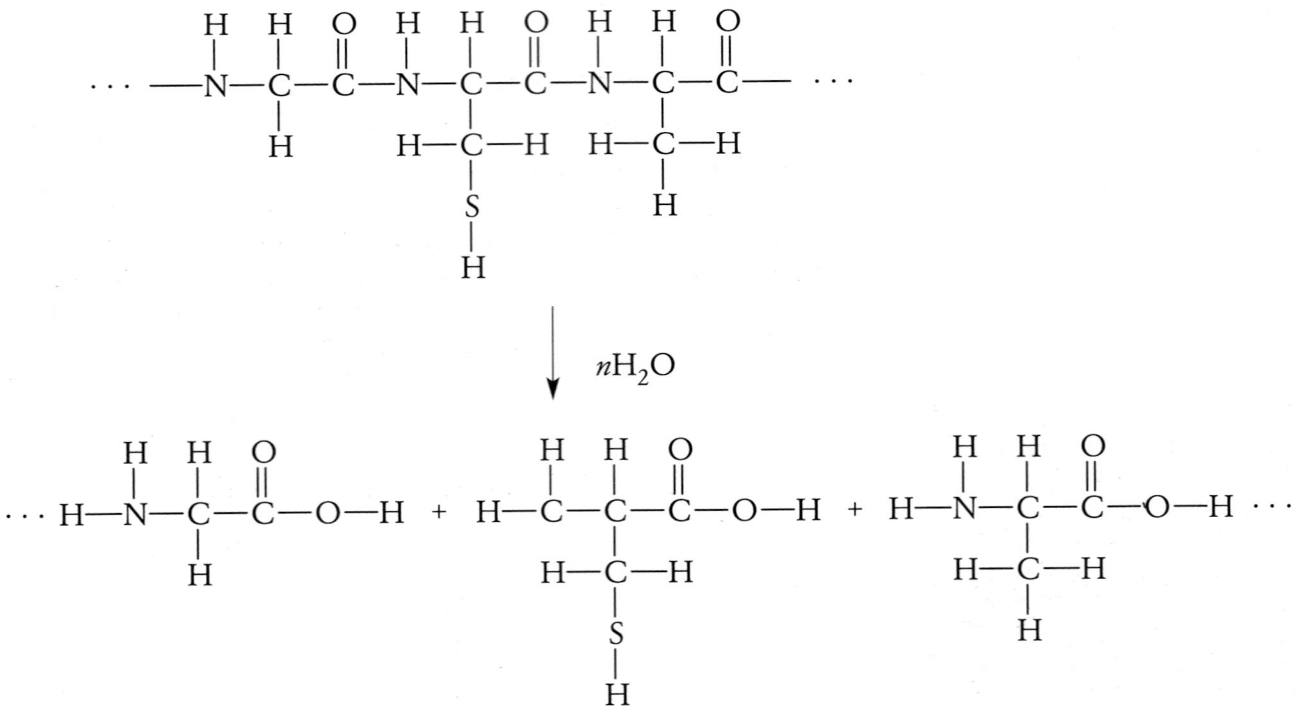

$$\downarrow \quad nH_2O$$

Fats and oils

Fats and oils in our diet

Edible **fats** and **oils** in our diet supply the body with energy, like carbohydrates, but they are a more concentrated source. Although it is released more slowly, fats and oils produce about twice as much energy as an equal mass of carbohydrate. Foods with a high fat or oil content include butter, margarine, cheese, cream and nuts.

Classifying fats and oils

Fats and oils can be of animal or vegetable origin. Some of them are shown in the Table 11.5.

What are fats and oils?

Fats and oils are esters that are made from a variety of quite large carboxylic acids called **fatty acids**. These fatty acids possess one carboxyl group which is attached to a long hydrocarbon chain. The fatty acid molecules all join with **glycerol** to make the esters present in fats and oils. Glycerol is not a simple alkanol because each glycerol molecule has three –OH groups attached to it, making it a triol. Three ways of writing structures for glycerol are shown below.

Animal	Vegetable
beef fat	olive oil
pork fat	sunflower oil
chicken fat	rapeseed oil
cod liver oil	
halibut oil	
whale oil	

Table 11.5 ▲
Classification of some fats and oils.

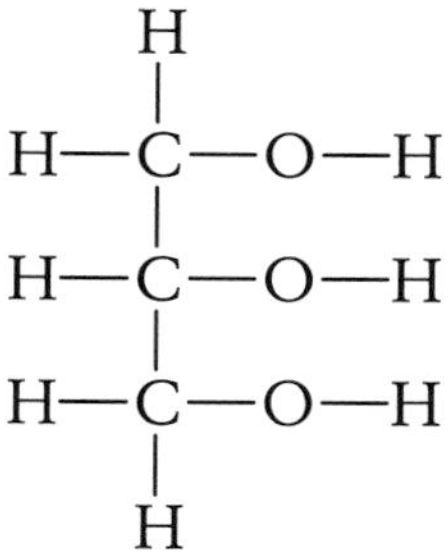

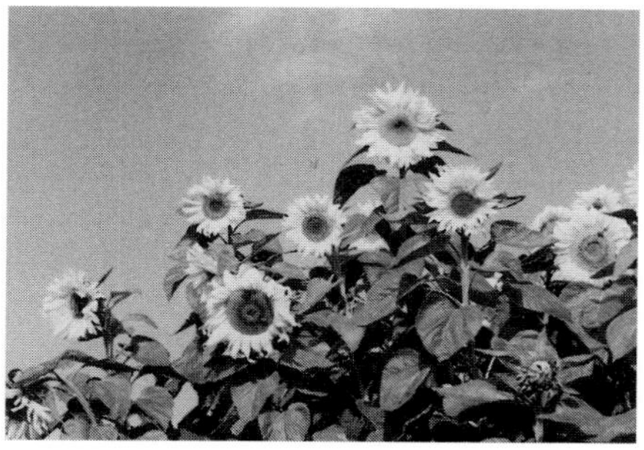

Figure 11.12 ▲
Sunflowers are grown in many parts of the world for the oil that can be extracted from their seeds.

$$CH_2OHCHOHCH_2OH \quad or \quad \begin{array}{c} CH_2OH \\ | \\ CHOH \\ | \\ CH_2OH \end{array}$$

or

$$\begin{array}{c} H \\ | \\ H-C-O-H \\ | \\ H-C-O-H \\ | \\ H-C-O-H \\ | \\ H \end{array}$$

The systematic name for glycerol is propane-1,2,3-triol. This is because the parent alkane is propane and there are three –OH groups present, one on each of the three carbon atoms in the molecule.

When a fat or oil forms, each glycerol molecule can make ester links with three fatty acid molecules. In the equation shown below, a fatty acid is represented by the formula RCOOH, where R is a large alkyl group.

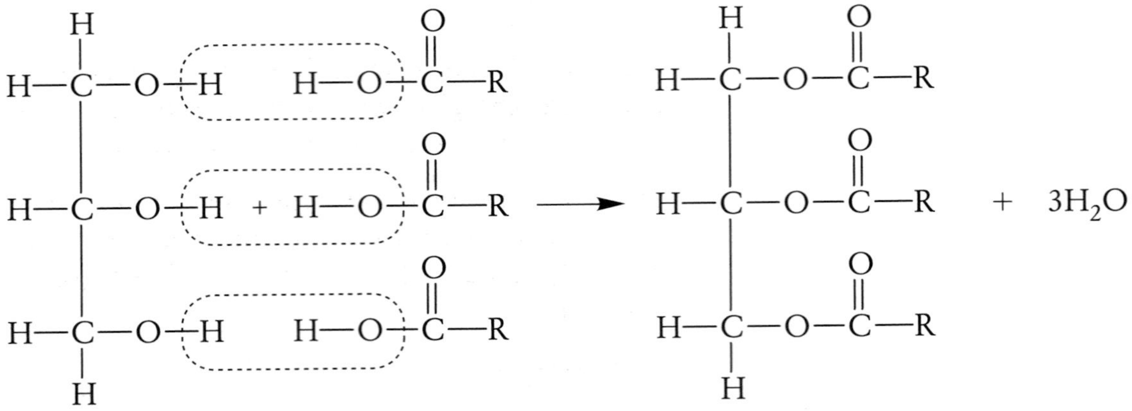

glycerol   fatty acid   fat or oil

Using shortened structural formulae, the same equation can be written as follows:

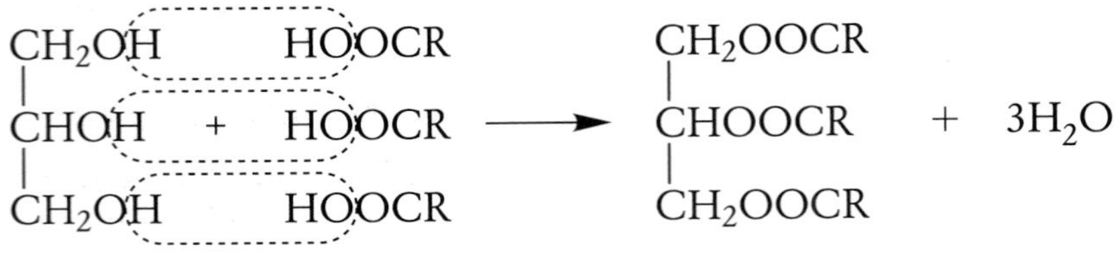

glycerol   fatty acid   fat or oil

Fats and oils in which all three fatty acid molecules are the same are not very common. More usually three different fatty acid molecules form ester links with glycerol. As a result, most fats and oils have the following structure:

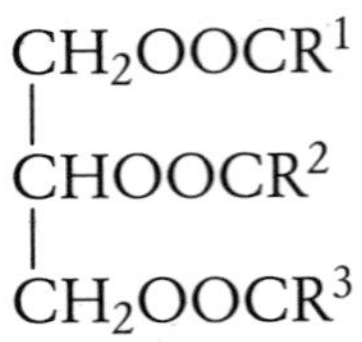

$$CH_2OOCR^1$$
$$CHOOCR^2$$
$$CH_2OOCR^3$$

R^1, R^2 and R^3 are different alkyl groups.

Fatty acids can be either saturated or unsaturated straight chain carboxylic acids. Some examples are given in Table 11.6.

Name	Molecular formula	Structural formula
palmitic acid	$C_{15}H_{31}COOH$	$CH_3(CH_2)_{14}COOH$
stearic acid	$C_{17}H_{35}COOH$	$CH_3(CH_2)_{16}COOH$
oleic acid	$C_{17}H_{33}COOH$	$CH_3(CH_2)_7CH=CH(CH_2)_7COOH$

Table 11.6 ▲
Some fatty acids.

Fats and oils can be hydrolysed using superheated steam to produce fatty acids and glycerol. The ratio of these products is always *three* moles of fatty acids to *one* mole of glycerol.

In a relatively rare case where only one fatty acid is present in an oil or fat molecule, the 3:1 ratio of fatty acid molecules to glycerol molecules produced on hydrolysis of an oil or fat can be seen in the equation below.

$$\begin{array}{l}
CH_2OOCC_{17}H_{35} \\
| \\
CHOOCC_{17}H_{35} \\
| \\
CH_2OOCC_{17}H_{35}
\end{array} + 3H_2O \longrightarrow \begin{array}{l}
CH_2OH \\
| \\
CHOH \\
| \\
CH_2OH
\end{array} + 3C_{17}H_{35}COOH$$

glycerol stearate   glycerol   stearic acid

Glycerol stearate is found in animal fat.

Differences between fats and oils

Both fats and oils are mixtures that contain both saturated and unsaturated hydrocarbon chains in their ester molecules. Oils have lower melting points than fats with a similar molecular size because they have a higher degree of unsaturation. The presence of C=C bonds makes it more difficult for the molecules to pack closely together. This reduces the forces of attraction between the molecules and causes the oils, which are more unsaturated, to have lower melting points than fats.

In the test for unsaturation, using bromine solution, oils decolourise more bromine solution than an equal mass of fat. The addition reaction that takes place is similar to that between alkenes and bromine.

$$\cdots \; -\overset{\displaystyle H}{\underset{\displaystyle |}{C}}=\overset{\displaystyle H}{\underset{\displaystyle |}{C}}- \; \cdots \; + \; Br-Br \longrightarrow \; \cdots \; -\overset{\displaystyle H}{\underset{\displaystyle |}{\underset{\displaystyle Br}{C}}}-\overset{\displaystyle H}{\underset{\displaystyle |}{\underset{\displaystyle Br}{C}}}- \; \cdots$$

part of a fat or oil molecule

Turning oils into fats

Margarines are made by passing hydrogen, under pressure, into hot vegetable oils in the presence of a catalyst. The hydrogen adds to some of the carbon-to-carbon double covalent bonds, C=C, in the fatty acid chains. This reduces the unsaturation resulting in an increase in melting point.

$$\ldots \ \begin{matrix} H & H \\ | & | \\ -C{=}C- \\ \end{matrix} \ \ldots \ + \ H_2 \ \longrightarrow \ \ldots \ \begin{matrix} H & H \\ | & | \\ -C{-}C- \\ | & | \\ H & H \end{matrix} \ \ldots$$

double bond in the
fatty acid chain of an oil

This addition reaction involving hydrogen is referred to as the 'hardening' of oils because a liquid oil is turned into a soft solid fat. Effectively, when for example sunflower oil is converted into sunflower margarine, the reaction that takes place may be summarised in the following way:

sunflower oil + hydrogen → sunflower margarine

Saturated fats and heart disease

There is evidence to suggest that there is a link between a high intake of saturated fat in the diet and heart disease. There is a very high incidence of heart disease in Scotland, particularly among men (see Figure 11.13). The situation could be improved if we were to lower our total fat intake and reduce the proportion of saturated fat in our diet. It is interesting to note that Mediterranean countries, such as France, Spain and Italy have a much lower incidence of heart disease. It seems likely that this relates to the greater proportion of unsaturated oils consumed by people in these countries.

Questions

17 Which are generally the more unsaturated – fats or oils?

18 Describe the test and result for unsaturation.

19 Fats and oils are esters.
  (a) Draw the ester link.
  (b) Name the products when a fat or oil is hydrolysed using superheated steam.

20 What illness is saturated fat associated with?

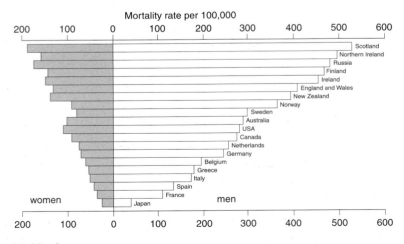

Figure 11.13 ▲

Coronary heart disease death rates for men (unshaded) and women (shaded) per million of population in Europe for the 40–69 age group in 1987.

In questions 1–4 choose the correct word **from the following list** to complete the sentence.

saturated, unsaturated, peptide, ester, addition, condensation, hydrolysis

1 When amino acids join to make protein molecules _____ links are formed.

2 The formation of starch from glucose is an example of _____ polymerisation.

3 Oils are generally more _____ than fats.

4 The formation of fatty acids and glycerol from oils or fats is an example of _____.

5 Which of the following does **not** give a positive result when tested with Benedict's solution?
   A Fructose
   B Glucose
   C Maltose
   D Sucrose

6 Which of the following correctly gives the elements that are present in all carbohydrates, fats and proteins?

	Carbohydrates	Fats	Proteins
A	C, H, O	C, H, O, N	C, H, O
B	C, H, O, N	C, H, O	C, H, O
C	C, H, O	C, H, O	C, H, O, N
D	C, H, O, N	C, H, O, N	C, H, O

7 Which of the following pairs of carbohydrates both have molecular formula $C_6H_{12}O_6$?
   A Fructose and glucose
   B Glucose and maltose
   C Fructose and sucrose
   D Maltose and sucrose

8 Which two terms can be used to describe insulin?
   A Addition polymer and hormone
   B Condensation polymer and carbohydrate
   C Addition polymer and protein
   D Condensation polymer and hormone

9 Edible fats and oils are mixtures of
   A carboxylic acids
   B alkanols
   C hydrocarbons
   D esters

10 (a) What is the ratio of hydrogen to oxygen atoms in any carbohydrate molecule?
    (b) Starch can be hydrolysed using enzymes. What can be used in place of enzymes to bring about this reaction?
    (c) Why do body enzymes not work well at high temperatures?

11 Human body fat contains four main acids:

Fatty acid	Molecular formula	%
Oleic acid	$C_{17}H_{33}COOH$	47
Palmitic acid	$C_{15}H_{31}COOH$	24
Linoleic acid	$C_{17}H_{31}COOH$	10
Stearic acid	$C_{17}H_{35}COOH$	8

    (a) Draw a bar graph to show the percentages of each fatty acid in human body fat.
    (b) Which of the above acids is/are saturated?

12 During margarine manufacture, some unsaturation is removed from vegetable oils. How is this achieved?

13 Part of a protein molecule is shown below.

$$\cdots -\underset{\underset{\displaystyle \;}{|}}{N} - \underset{\underset{\displaystyle CH}{|}}{C} - \underset{}{\overset{O}{\underset{}{C}}} - \underset{}{N} - \underset{\underset{\displaystyle CHOH}{|}}{C} - \overset{O}{C} - N - \underset{\underset{\displaystyle CH_2OH}{|}}{C} - \overset{O}{C} - \cdots$$

(H H O H H O H H O arranged above the chain, CH₃ CH₃ below first CH, CH₃ below CHOH)

    (a) Draw a structural formula for one of the amino acids that could be obtained on hydrolysis of the protein.
    (b) Draw a formula for the amine group.
    (c) What type of polymerisation takes place when amino acid molecules join together in proteins.

14 (a) When a fat is hydrolysed, what is the ratio of glycerol molecules to fatty acid molecules obtained?
    (b) Draw a full structural formula for glycerol.

Acids, Bases and Metals

The pH scale

It is not possible to tell, just by looking at it, whether an acid is strong, like sulphuric acid in a car battery, or weak, like ethanoic acid in vinegar. Chemists measure the difference between strong and weak acids using a scale of numbers called the **pH scale**. The same scale can be used to give information about the differing strengths of alkalis. For example, sodium hydroxide (caustic soda) is a strong alkali, whereas sodium hydrogencarbonate (bicarbonate of soda) is a weak alkali.

The pH scale is a continuous range from below 0 to above 14 (see Figure 12.1)

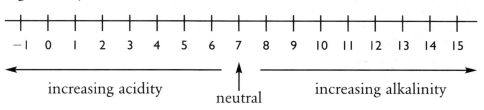

Figure 12.1 ▲

How to measure pH

An easy way of finding out the approximate pH of a solution is either to dip pH paper into it, or to add a little pH indicator solution to it. The colours produced are compared with those on standard colour charts in order to obtain the approximate pH value. pH indicators, or universal indicators as they are sometimes called, can be made so that they follow the colours of the rainbow. Red, orange and yellow are 'acidic colours', green is the neutral colour and blue, indigo and violet are 'alkaline colours'. Red indicates that a solution is strongly acidic and violet indicates a strongly alkaline solution.

A more accurate method of determining the pH of a solution is to place the probe of a pH meter into it. pH meters are capable of giving a pH value accurate to one decimal place or better. For example, the pH of a fizzy drink might be found to be between 2 and 3 using pH paper or pH solution. However, when tested with a pH meter, the more accurate pH value could be 2.4, for example (see Figure 12.2).

What the pH numbers mean

Any solution which has a pH of **less than 7** is **acidic**. The *lower* the number the stronger the acid. For example, sulphuric acid from a car battery, has a pH of less than 1, whereas vinegar, which is a solution of ethanoic acid in water, has a pH of about 3. Battery acid is more acidic than vinegar.

Figure 12.2 ▲
Two methods of measuring pH: with pH paper (top) and with a pH meter (bottom).

Question

1 Ask your teacher for some pH paper to test some household substances. Make a solution of any solid you wish to test because dry solids do not affect pH paper. Present your results as a table.

Solutions with a pH **greater than 7** are **alkaline**. In this case the *higher* the number the stronger the alkali. A solution of caustic soda, for cleaning drains, has a pH of more than 14, whereas a solution of bicarbonate of soda has a pH of about 9. Caustic soda is more alkaline than bicarbonate of soda.

Solutions that have a pH of 7 are said to be **neutral**. For example, a solution of common salt, sodium chloride, is neutral. Water itself has a pH of 7 and is therefore also described as neutral.

Water – mainly molecular, but slightly ionic!

Water is an unusual liquid. If you use a low voltage and a normal ammeter it appears to be a non-conductor and therefore exists as molecules. However, if a very sensitive ammeter is used, it can be seen that water *is* a conductor. Since water conducts slightly there must be some ions present. Chemists believe that in pure water, and in all aqueous solutions, a **reversible reaction** takes place. In this reversible reaction, water molecules are constantly breaking up into hydrogen and hydroxide ions, which in turn are constantly reacting to reform water molecules. The equation for this is as follows:

$$H_2O(l) \rightleftharpoons H^+(aq) + OH^-(aq)$$

An **equilibrium** is set up when the rate at which water molecules break up to form ions is equal to the rate at which the ions join to form water molecules. At equilibrium in this reaction there is a much greater concentration of molecules than ions. As a result water is a very poor conductor of electricity because it is ions that carry the current. It is important to understand that when a reversible reaction is at equilibrium, the concentrations of reactants and products remain *constant*, but they are not necessarily equal.

Each water molecule is capable of producing one hydrogen ion and one hydroxide ion. Therefore pure water must contain *equal* concentrations of hydrogen and hydroxide ions. Also, because water is such a poor conductor, we can further say that pure water must contain *tiny* but equal concentrations of hydrogen ions, $H^+(aq)$, and hydroxide ions, $OH^-(aq)$.

Questions

2 Draw a labelled diagram of the apparatus you could use to show that water conducts electricity.

3 Electrons carry electricity through metals. What particles carry electricity through water?

Name and formula	Ions in solution
hydrochloric acid HCl	$H^+(aq) + Cl^-(aq)$
nitric acid HNO_3	$H^+(aq) + NO_3^-(aq)$
sulphuric acid H_2SO_4	$2H^+(aq) + SO_4^{2-}(aq)$

Table 12.1 ▲
Common laboratory acids.

Acids, alkalis and ions

Acids are compounds that produce hydrogen ions, $H^+(aq)$, when they dissolve in water. Pure acids are covalent, but they break up into ions when dissolved in water. For example, hydrochloric acid is formed when hydrogen chloride gas dissolves in water and breaks up to produce hydrogen ions and chloride ions.

$$HCl(g) + aq \rightarrow H^+(aq) + Cl^-(aq)$$

Since acids dissolve in water to produce hydrogen ions, solutions of acids have a higher concentration of hydrogen ions than hydroxide ions. This is because pure water itself contains equal concentrations of hydrogen and hydroxide ions.

The other common laboratory acids, nitric acid and sulphuric acid, also break up in water to produce hydrogen ions. Nitric acid also produces nitrate ions and sulphuric acid also produces sulphate ions.

$$HNO_3(l) + aq \rightarrow H^+(aq) + NO_3^-(aq)$$

$$H_2SO_4(l) + aq \rightarrow 2H^+(aq) + SO_4^{2-}(aq)$$

When an acid solution is diluted by the addition of water, the concentration of $H^+(aq)$ ions decreases, making the solution less acidic. As a result the pH increases and moves towards 7.

Alkalis are compounds that produce hydroxide ions, $OH^-(aq)$, when they dissolve in water. For example, sodium hydroxide dissolves to produce sodium ions and hydroxide ions.

$$Na^+OH^-(s) + aq \rightarrow Na^+(aq) + OH^-(aq)$$

Since alkalis dissolve in water to produce hydroxide ions, solutions of alkalis have a higher concentration of hydroxide ions than hydrogen ions.

Two other common laboratory alkalis, potassium hydroxide and calcium hydroxide, also dissolve to produce hydroxide ions.

$$K^+OH^-(s) + aq \rightarrow K^+(aq) + OH^-(aq)$$

$$Ca^{2+}(OH^-)_2(s) + aq \rightarrow Ca^{2+}(aq) + 2OH^-(aq)$$

Calcium hydroxide solution is more commonly known as lime water.

Unlike pure acids, which are covalent, pure metal hydroxides are ionic. When a metal hydroxide dissolves in water, the ions in the solid lattice are simply freed from their positions.

An example of a covalent compound that dissolves in water to produce an alkaline solution is ammonia gas, NH_3.

Name and formula	Ions in solution
sodium hydroxide NaOH	$Na^+(aq) + OH^-(aq)$
potassium hydroxide KOH	$K^+(aq) + OH^-(aq)$
calcium hydroxide $Ca(OH)_2$	$Ca^{2+}(aq) + 2OH^-(aq)$

Table 12.2 ▲
Common laboratory alkalis.

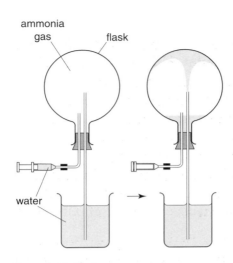

Figure 12.3 ▲
A fountain experiment.

Questions

4 When carrying out fountain experiments pH indicator is often added to the water in the beaker first. Predict any colour changes that might be expected to take place during a fountain experiment using:

(a) ammonia
(b) hydrogen chloride
(c) sulphur dioxide.

5 What name is usually given to a solution of hydrogen chloride in water?

Unlike hydrogen chloride, it reacts *reversibly* with water, producing ammonium ions and hydroxide ions.

$$NH_3(g) + H_2O(l) \rightleftharpoons NH_4^+(aq) + OH^-(aq)$$

When an alkaline solution is diluted by the addition of water, the concentration of OH^-(aq) ions decreases and the solution becomes less alkaline. As a result the pH decreases and moves towards 7.

A fountain experiment

This is an interesting way of showing that some gases are very soluble in water. For example, a little water is injected into a flask that contains ammonia gas as in Figure 12.3. Almost all of the ammonia dissolves in the injected water and, as a result, the pressure in the flask drops. Atmospheric pressure then pushes water from the beaker below up into the flask. This happens with such force that the in-coming water looks like a fountain. The experiment works because ammonia is very soluble in water, so most of the large volume of ammonia in the flask is able to dissolve in the small amount of water that is injected at the start.

Fountain experiments can be carried out with other very soluble gases such as hydrogen chloride and sulphur dioxide.

Non-metal oxides and acidity

Simple experiments show that non-metal oxides that dissolve in water produce *acidic* solutions. For example, the gases carbon dioxide, sulphur dioxide and nitrogen dioxide all dissolve to produce acids. The white solid phosphorus(V) oxide, that is produced when phosphorus burns in a plentiful supply of air, also dissolves to produce an acid.

Non-metal oxide	Acid produced
carbon dioxide, CO_2	carbonic acid, H_2CO_3
sulphur dioxide, SO_2	sulphurous acid, H_2SO_3
nitrogen dioxide, NO_2	nitric acid, HNO_3
phosphorus(V) oxide, P_2O_5	phosphoric acid, H_3PO_4

Table 12.3 ▲

An example of an insoluble, neutral non-metal oxide is carbon monoxide, CO. Water is also a non-metal oxide, being an oxide of hydrogen, H_2O. Water can be thought of as a neutral oxide because it produces equal concentrations of hydrogen and hydroxide ions, and because it has a pH of 7.

Questions

6 Write a balanced formula equation for the reaction between water and phosphorus(V) oxide to produce phosphoric acid.

7 Use a data booklet to find out which of the following metal oxides is soluble in water and then write a balanced formula equation for the reaction that takes place:
ZnO, NiO, CaO, FeO.

Metal oxides and alkalinity

All metal oxides and metal hydroxides that dissolve in water produce *alkaline* solutions. The oxides and hydroxides of the Group 1 metals are all very soluble in water, whereas in Group 2, solubility increases down the group from the (almost) insoluble magnesium oxide and magnesium hydroxide, to the very soluble corresponding barium compounds. The oxides and hydroxides of all other metals tend to be very insoluble and do not affect the pH of water. Examples of very insoluble metal oxides include copper(II) oxide, CuO, and iron(III) oxide, Fe_2O_3. Examples of very insoluble metal hydroxides include copper(II) hydroxide, $Cu(OH)_2$, and aluminium hydroxide, $Al(OH)_3$.

In the case of soluble metal oxides, the metal oxide reacts with water to form the hydroxide. For example sodium oxide reacts with water to form sodium hydroxide.

$$Na_2O(s) + H_2O(l) \rightarrow 2NaOH(aq)$$

Magnesium oxide reacts with water to produce magnesium hydroxide which, although it has a solubility of less than 1 gram per litre, is still sufficiently alkaline to turn pH indicator solution violet.

$$MgO(s) + H_2O(l) \rightarrow Mg(OH)_2(aq)$$

A mixture of magnesium hydroxide and water is sold as 'milk of magnesia', which is used as a cure for acid indigestion.

Acids and alkalis in the home

Many of the things that we find in our homes contain either acids or alkalis. For example, acids in drinks give them a pleasant 'sharp' or sour taste. Carbon dioxide is used to make drinks fizzy, but it also lowers the pH and makes them acidic due to the formation of carbonic acid. Citric acid is present in all citrus fruits like lemons, oranges, grapefruits and limes. Phosphoric acid is used for making cola drinks. Vinegar has a sour taste due to the presence of ethanoic acid, which is also used in the preparation of certain sauces. Vinegar is also used for the pickling of vegetables such as beetroot. (See Figure 12.4 and Table 12.4.)

Acid in the home	Present in
carbonic acid	fizzy drinks
citric acid	citrus fruits, e.g. lemons
ethanoic acid	vinegar, sauces, pickles
phosphoric acid	cola drinks

Table 12.4 ▲
Some common household acids.

Figure 12.4 ▲
All of these contain acids.

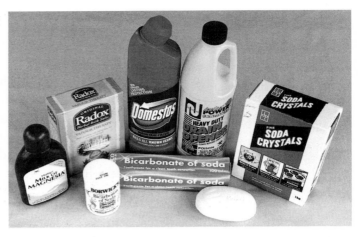

Figure 12.5 ▲
All of these contain alkalis.

Question

8 A pupil obtained the following pH values after testing some household substances with indicator paper.

Vinegar 3; lemon juice 2; cola 2.5; eyewash 5; washing soda 11; toothpaste 8; tea 6; oven cleaner 13.

Present this information another way with the aid of a pH scale.

Alkalis do not make drinks or food taste better – think of the taste of bicarbonate of soda or milk of magnesia, both of which are alkaline. However, there are important uses for alkalis in the home – mostly concerned with cleaning. Some oven cleaners and bleaches contain sodium hydroxide. Drain cleaner is pure sodium hydroxide. Sodium carbonate is sold as washing soda and is present in dishwasher powder and tablets. Bicarbonate of soda is pure sodium hydrogencarbonate and baking soda contains the same alkali.

Alkali in the home	Present in
sodium carbonate	washing soda, dishwasher powder
sodium hydrogencarbonate	bicarbonate of soda and baking soda
sodium hydroxide	oven and drain cleaner, bleach
magnesium hydroxide	milk of magnesia

Table 12.5 ▲
Some common household alkalis.

Concentration and the mole

Chemists give the concentration of a solution as the number of moles of solute present in one litre of solution. In other words, the units of concentration are moles per litre or $mol\,l^{-1}$. For example, a $2\,mol\,l^{-1}$ solution of sodium hydroxide contains 2 moles of solute, sodium hydroxide, in 1 litre of solution.

The relationship between concentration of solution (C), number of moles of solute (n) and volume of solution (V) is:

$$\text{Concentration of solution (in } mol\,l^{-1}) = \frac{\text{number of moles of solute}}{\text{volume of solution (in litres)}}$$

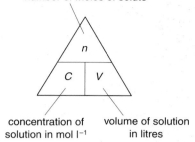

number of moles of solute

n

C | V

concentration of solution in mol l^{-1}

volume of solution in litres

mass in grams

m

n | FM

number of moles

formula mass

The relevant triangle of knowledge is:

$$n = C \times V$$

$$C = \frac{n}{V}$$

$$V = \frac{n}{C}$$

Although chemists express concentration in terms of *moles* of solute per litre of solution, when weighing out the solute they note the *mass* in *grams*. Therefore, before calculating the concentration, there is a need to convert grams into moles. This can be done using the triangle of knowledge from page 72 which relates mass of substance (*m*), number of moles of substance (*n*) and formula mass (*FM*).

$$n = \frac{m}{FM}$$

Questions

9 Calculate the concentration of a glucose solution, given that 0.2 moles of glucose are dissolved in water and made up to 250 cm^3 of solution.

10 Calculate the concentration of a solution of calcium hydroxide, Ca(OH)$_2$, given that it contains 1.48 g of Ca(OH)$_2$ in 800 cm^3 of solution.

11 Calculate the volume of solution produced if 22.5 g of oxalic acid, (COOH)$_2$, is used to make a solution with a concentration of 0.05 mol l^{-1}.

Example 1
Calculate the concentration of a solution of potassium nitrate given that 2 moles of potassium nitrate are dissolved in water and made up to a final volume of 4 litres.

Concentration of potassium nitrate, $C = \dfrac{n}{V} = \dfrac{2}{4} = 0.5\,\text{mol}\,l^{-1}$

Example 2
Calculate the concentration (in mol l^{-1}) of a solution of sodium hydroxide, NaOH, given that it contains 24 g of NaOH in 250 cm^3 of solution.

Number of moles of NaOH, $n = \dfrac{m}{FM} = \dfrac{24}{40} = 0.6$

Concentration of NaOH, $C = \dfrac{n}{V} = \dfrac{0.6}{0.25} = 2.4\,\text{mol}\,l^{-1}$

Note: 250 cm^3 = 0.25 litres

Example 3
Calculate the volume of solution produced if 50.5 g of potassium nitrate, KNO$_3$, is used to make a solution with a concentration of 0.2 mol l^{-1}.

Number of moles of KNO$_3$, $n = \dfrac{m}{FM} = \dfrac{50.5}{101} = 0.5$

Volume of KNO$_3$ solution, $V = \dfrac{n}{C} = \dfrac{0.5}{0.2} = 2.5\,\text{litres}$

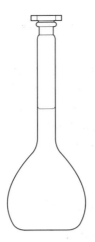

Figure 12.6 ▲
A standard flask.

Preparation of a standard solution

For chemists, it is very important to know the exact concentration of a solution if calculations involving that solution are to be carried out. If the concentration of a solution is known accurately it is then referred to as a **standard solution**.

Compounds that are used to make standard solutions must be available in a pure state. This is because, when the compound is weighed out, it is important that the mass to be used in calculations refers only to that compound and not to any impurities.

Sodium carbonate, for example, is a compound that is available in a pure state and can therefore be used for the preparation of a standard solution.

From an accurately known mass of a pure compound the number of moles present can be calculated accurately. However, in order that the concentration of a solution is known accurately, an accurately known number of moles of solute must be present in an accurately known volume of solution. In order that the volume of a solution is known accurately, chemists use a *standard flask* (see Figure 12.6) which has a mark around its narrow neck showing the correct level for the solution.

Aim
To prepare $250 \, cm^3$ of a $0.10 \, mol \, l^{-1}$ solution of sodium carbonate, Na_2CO_3.

Calculation
Number of moles of Na_2CO_3 required, $n = C \times V = 0.10 \times 0.25$
$$= 0.025$$

Mass of Na_2CO_3 required, $m = n \times FM = 0.025 \times 106 = 2.65 \, g$

Procedure
1 Weight accurately $2.65 \, g$ of Na_2CO_3 in a small beaker.

2 Add a small volume of deionised (or distilled) water and stir to dissolve.

3 Transfer to a $250 \, cm^3$ standard flask along with washings from the beaker.

4 Make up to the mark carefully using more deionised (or distilled) water, using a dropper for the last few drops.

5 Invert several times to mix thoroughly.

Concentrated and dilute acids and alkalis

Concentrated acids and alkalis contain a lot of acid or alkali in a small amount of water. For example, the concentrated sulphuric acid purchased from suppliers for use in school chemistry departments consists of 98% acid and only 2% water.

Questions

12 Calculate the mass of sodium chloride that would be required to produce $500 \, cm^3$ of a solution with a concentration of $0.2 \, mol \, l^{-1}$.

13 Calculate the mass in grams of solute present in each of the following solutions:

(a) 2 litres of $2 \, mol \, l^{-1}$ KOH

(b) $250 \, cm^3$ of $0.1 \, mol \, l^{-1}$ $(NH_4)_2SO_4$

Dilute acids and alkalis have a small amount of acid or alkali in a lot of water.

When expressed in the units $mol\,l^{-1}$, concentrated acids and alkalis usually have concentrations of $10\,mol\,l^{-1}$ or more. Concentrated hydrochloric acid from suppliers has a concentration of about $11\,mol\,l^{-1}$, whereas 98% sulphuric acid has a concentration of about $18\,mol\,l^{-1}$. Dilute acids and alkalis, as used in school laboratories, generally have a concentration of $2\,mol\,l^{-1}$ or less.

Strong and weak acids and alkalis

A **strong acid** is one that breaks up *completely* in aqueous solution to produce ions. Chemists often use the word 'dissociates' to mean 'breaks up' in this context. The three common laboratory acids, hydrochloric, nitric and sulphuric, all dissociate (break up) completely in aqueous solution to produce ions.

$$HCl(aq) \rightarrow H^+(aq) + Cl^-(aq)$$

$$HNO_3(aq) \rightarrow H^+(aq) + NO_3^-(aq)$$

$$H_2SO_4(aq) \rightarrow 2H^+(aq) + SO_4^{2-}(aq)$$

A **weak acid** is one that dissociates (breaks up) only *partially* in aqueous solution to produce ions. For example, all acids containing the carboxyl group are weak, including ethanoic acid.

$$CH_3COOH(aq) \rightleftharpoons CH_3COO^-(aq) + H^+(aq)$$

Equimolar solutions (solutions of equal concentrations in terms of moles of solute per litre) of strong and weak acids differ in pH, conductivity and rate of reaction with metal or carbonate. A typical results table would be as follows:

Test	$1\,mol\,l^{-1}$ HCl	$1\,mol\,l^{-1}$ CH_3COOH
pH	very low	low (below 7)
Conductivity	high	low
Reaction with Mg	fast	slow
Reaction with $CaCO_3$	fast	slow

Table 12.6 ▲

The pH of hydrochloric acid is lower than that of ethanoic acid because it has a higher concentration of $H^+(aq)$ ions. This also explains why the reactions with both magnesium and calcium carbonate are faster with hydrochloric acid. It is $H^+(aq)$ that reacts, both with the metal and with the carbonate. The conductivity of hydrochloric acid is higher than that of ethanoic acid due to a greater concentration of ions, both positive and negative, since both types carry the current.

A **strong alkali** is one that dissociates (breaks up) *completely* in aqueous solution to produce ions. Soluble metal hydroxides, such as those from group 1 of the Periodic Table, are examples of strong alkalis. Unlike acids, which exist as polar covalent molecules when pure, these alkalis are ionic solids. Dissolving simply frees the ions in the lattice. Sodium, potassium and calcium hydroxide are examples of strong alkalis.

$$Na^+OH^-(s) + aq \rightarrow Na^+(aq) + OH^-(aq)$$

$$K^+OH^-(s) + aq \rightarrow K^+(aq) + OH^-(aq)$$

$$Ca^{2+}(OH^-)_2(s) + aq \rightarrow Ca^{2+}(aq) + 2OH^-(aq)$$

A **weak alkali** is one that dissociates *partially* in aqueous solution to produce ions. Ammonia solution is an example of a weak alkali.

$$NH_3(aq) + H_2O(l) \rightleftharpoons NH_4^+(aq) + OH^-(aq)$$

Equimolar solutions of strong and weak alkalis differ in pH and conductivity. A typical results table would be as follows:

Test	$1\ mol\,l^{-1}$ NaOH	$1\ mol\,l^{-1}$ NH$_3$
pH	very high	high (above 7)
Conductivity	high	low

Table 12.7 ▲

The pH of the sodium hydroxide solution is higher than that of the ammonia solution because it has a higher concentration of $OH^-(aq)$ ions. The conductivity of the sodium hydroxide solution is also higher than that of the ammonia solution due to a greater concentration of ions, both positive and negative, since both types carry the current.

Strength and **concentration** are terms that are often used incorrectly. They do *not* have the same chemical meaning. Strength is used to describe the extent of dissociation into ions. Concentration is used to describe the number of moles of solute in a certain volume of solution.

Questions

14 (a) State what is meant by a strong acid.
   (b) Give an example of a strong acid.

15 (a) State what is meant by a weak alkali.
   (b) Give a example of a weak alkali.

16 Draw a labelled diagram of the apparatus that you would use to measure the conductivity of dilute solutions of acids or alkalis.

17 Explain why a dilute alkali is not necessarily weak.

In questions 1–4 choose the correct word **from the following list** to complete the sentence.

molecules, ions, upwards, downwards, acidic, alkaline, more, less

1 Acidic solutions have a pH that is _____ than 7.

2 Metal oxides that dissolve in water produce _____ solutions.

3 Strong alkalis dissociate completely in aqueous solution to produce _____.

4 Diluting an acidic solution with water moves the pH _____ towards 7.

5 Which of the following oxides dissolves in water to give an acidic solution?
A PbO_2
B SnO_2
C SiO_2
D CO_2

6 A solution in which 20 g of sodium hydroxide is present in 250 cm^3 of solution has a concentration in mol l^{-1} of:
A 0.5   B 1   C 2   D 4

7 Which of the following is the only common alkaline gas?
A Hydrogen
B Ammonia
C Nitrogen
D Hydrogen chloride

8 Which of the following shows the correct ascending order for the pH of 0.1 mol l^{-1} solutions of the compounds referred to?
A $NaOH$, NH_3, CH_3COOH, HNO_3
B NH_3, $NaOH$, HNO_3, CH_3COOH
C CH_3COOH, HNO_3, $NaOH$, NH_3
D HNO_3, CH_3COOH, NH_3, $NaOH$

9 A bottle is labelled 10 mol l^{-1} ethanoic acid solution – corrosive. This can best be described as:
A a concentrated solution of a weak acid
B a concentrated solution of a strong acid
C a dilute solution of a weak acid
D a dilute solution of a strong acid.

10 Pure water conducts electricity very slightly because of a reversible reaction that takes place between water molecules.

(a) Write a formula equation for this reaction.
(b) Why is the reaction described as 'reversible'?

11 Dilute sulphuric acid is used in many school experiments.

(a) Write formulae for the two ions it produces in aqueous solution.
(b) What is meant by the word 'acid'?

12 Calcium hydroxide solution is also known as lime water.

(a) Write an ionic formula for calcium hydroxide.
(b) What happens to the pH of calcium hydroxide solution as it is diluted with water?

13 Ammonia is very soluble in water producing a weak alkaline solution.

(a) What is meant by the term 'a weak alkali'?
(b) Write a formula equation for the reaction of ammonia with water.
(c) Name a gas that dissolves to give an acidic solution.

14 In an investigation into the ability of different 1 mol l^{-1} solutions to conduct electricity, the following results were obtained.

Solution	Current/ mA
Sodium chloride , $Na^+(aq) + Cl^-(aq)$	30
Sodium hydroxide, $Na^+(aq) + OH^-(aq)$	50
Hydrochloric acid, $H^+(aq) + Cl^-(aq)$	90

(a) Plot a bar graph to show these results.
(b) Different ions have different abilities to carry an electric current. Based on the results shown which ion is best at carrying an electric current?

15 100 cm^3 of 0.01 mol l^{-1} lead(II) nitrate, $Pb(NO_3)_2$, was required for an experiment.

(a) Calculate the mass of lead(II) nitrate needed.
(b) State how you would make this solution.

13 Salt Preparation

Reactions of acids

Acids combine readily with a variety of compounds in reactions where the pH of the solution moves towards 7. These are **neutralisation** reactions, during which the $H^+(aq)$ ions from the acid are removed, reducing the acidity. Compounds that react with the $H^+(aq)$ ions from acids to produce water are called **bases**. Examples of bases include metal oxides, hydroxides, carbonates, hydrogencarbonates and ammonia.

Alkalis are a sub-group of the group of bases, being those bases that dissolve in water producing alkaline solutions. The alkali metal hydroxides, such as sodium hydroxide, and ammonia are examples of alkalis.

It is worth noting that the neutralisation of a solution of an acid moves the pH *up* towards 7, but the neutralisation of a solution of an alkali moves the pH *down* towards 7.

Salts

Salts are ionic compounds that are formed in neutralisation reactions between acids and bases and between acids and metals. During these reactions the hydrogen ions of the acid are replaced by metal ions or ammonium ions as a salt is formed. Salts therefore contain a metal ion, or an ammonium ion, from the base or metal and a negative ion from the acid. The name of the salt is derived from the names of these positive and negative ions present in it. For example, sodium hydroxide produces *sodium* salts and hydrochloric acid (which contains the chloride ion) produces *chloride* salts. Tables 13.1 and 13.2 give the names of some salts produced by acids and bases or metals.

Base/metal	Positive ion/salt name
sodium hydroxide	sodium ...
copper(II) oxide	copper(II) ...
calcium carbonate	calcium ...
ammonia	ammonium ...
magnesium	magnesium ...

Table 13.1 ▲
Naming salts from bases/metals.

Acid	Negative ion/salt name
hydrochloric acid	... chloride
nitric acid	... nitrate
sulphuric acid	... sulphate
ethanoic acid	... ethanoate

Table 13.2 ▲
Naming salts from acids.

Alkali/acid reactions

When an alkali and an acid react, the products are a salt and water.

alkali + acid → salt + water

For example, sodium hydroxide solution reacts with hydrochloric acid to produce sodium chloride and water.

Word equation:

sodium + hydrochloric → sodium + water
hydroxide   acid   chloride

Balanced formula equation:

$$NaOH(aq) + HCl(aq) \rightarrow NaCl(aq) + H_2O(l)$$

Ionic equation:

$$Na^+(aq) + OH^-(aq) + H^+(aq) + Cl^-(aq) \rightarrow Na^+(aq) + Cl^-(aq) + H_2O(l)$$

(Notice how the ions are 'separated' if an ionic compound dissolves in water.)

Removing **spectator ions** (those which appear on both sides of the equation and do not react):

$$OH^-(aq) + H^+(aq) \rightarrow H_2O(l)$$

This last reaction is common to any reaction between solutions of an alkali and an acid. Hydrogen ions from the acid react with hydroxide ions from the alkali to form water.

Metal oxide/acid reactions

Just as with alkali/acid reactions, metal oxides also react with acids to give a salt and water.

metal oxide + acid → salt + water

For example, zinc oxide reacts with sulphuric acid to produce zinc sulphate and water.

Question

1 Write word, balanced formula and ionic equations for the reaction between potassium hydroxide solution and dilute nitric acid.

Word equation:

$$\text{zinc oxide} + \text{sulphuric acid} \rightarrow \text{zinc sulphate} + \text{water}$$

Balanced formula equation:

$$ZnO(s) + H_2SO_4(aq) \rightarrow ZnSO_4(aq) + H_2O(l)$$

Ionic equation:

$$Zn^{2+}O^{2-}(s) + 2H^+(aq) + SO_4^{2-}(aq) \rightarrow$$
$$Zn^{2+}(aq) + SO_4^{2-}(aq) + H_2O(l)$$

Removing the spectator ions gives:

$$O^{2-}(s) + 2H^+(aq) \rightarrow H_2O(l)$$

When acids and metal oxides react, hydrogen ions and oxide ions join to form water.

It should be noted that the charge on a zinc ion *cannot* be worked out by reference to the Periodic Table since it is found among the transition metals. Fortunately, in all of its compounds it is found as the zinc(II) ion, Zn^{2+}.

Question

2   Write word, balanced formula and ionic equations for the reaction between copper(II) oxide and dilute hydrochloric acid.

Metal carbonate/acid reactions

Metal carbonates (and hydrogencarbonates) react with acids to produce a salt, along with water and carbon dioxide gas.

metal carbonate + acid → salt + water + carbon dioxide

For example, when nitric acid is added to a solution of sodium carbonate in water, sodium nitrate, water and carbon dioxide are formed.

Word equation:

sodium carbonate + nitric acid → sodium nitrate + water + carbon dioxide

Balanced formula equation:

$$Na_2CO_3(aq) + 2HNO_3(aq) \rightarrow 2NaNO_3(aq) + H_2O(l) + CO_2(g)$$

Ionic equation:

$$2Na^+(aq) + CO_3^{2-}(aq) + 2H^+(aq) + 2NO_3^-(aq) \rightarrow$$
$$2Na^+(aq) + 2NO_3^-(aq) + H_2O(l) + CO_2(g)$$

Removing the spectator ions gives:

$$CO_3^{2-}(aq) + 2H^+(aq) \rightarrow H_2O(l) + CO_2(g)$$

When acids and carbonates react, hydrogen ions and carbonate ions combine to form water and carbon dioxide.

Sweet manufacturers make use of an acid/carbonate reaction when you bite into a sherbet lemon (see Figure 13.1).

Figure 13.1 ▲
Sherbet lemons contain bicarbonate of soda and citric acid. (Which is in excess?).

Question

3   Write word, balanced formula and ionic equations for the reaction between calcium carbonate and dilute hydrochloric acid. (Assume that calcium carbonate is insoluble in water.)

Metal/acid reactions

As a general rule, metals react with acids to give a salt and hydrogen gas, but there are exceptions. Some metals do not react at all and some form other products.

metal + acid → salt + hydrogen

For example, magnesium reacts with sulphuric acid to produce magnesium sulphate and hydrogen.

Word equation:

   magnesium + sulphuric acid → magnesium sulphate + hydrogen

Balanced formula equation:

$$Mg(s) + H_2SO_4(aq) \rightarrow MgSO_4(aq) + H_2(g)$$

Ionic equation:

$$Mg(s) + 2H^+(aq) + SO_4^{2-}(aq) \rightarrow Mg^{2+}(aq) + SO_4^{2-}(aq) + H_2(g)$$

Removing spectator ions gives:

$$Mg(s) + 2H^+(aq) \rightarrow Mg^{2+}(aq) + H_2(g)$$

When acids and metals react in this way, hydrogen ions are turned into hydrogen molecules.

Question

4   Write word, balanced formula and ionic equations for the reaction between zinc and dilute hydrochloric acid.

Test for hydrogen gas

When mixed with air, and a lighted taper is applied, hydrogen burns with a pop.

What is acid rain?

Rain water is naturally slightly acidic due to the presence of dissolved carbon dioxide which produces carbonic acid (Chapter 7). Since carbonic acid is very weak it does not lower the pH very much and does not harm living things. Rain water containing only carbonic acid is not considered to be 'acid rain'.

Unfortunately, when fossil fuels, particularly coal, are burned sulphur, which is present as an impurity, burns to produce sulphur dioxide (see Figure 13.2).

$$S(s) + O_2(g) \rightarrow SO_2(g)$$

Sulphur dioxide is a colourless, poisonous gas with a sharp, choking smell. It dissolves in rain water to produce an acid called sulphurous acid. This is not as weak as carbonic acid and so causes a lower pH. However, after a while the sulphurous acid changes into the strong acid, sulphuric acid, and the pH falls still lower. The rain water now has a much lower than normal pH and is described as '**acid rain**'.

Another gas that contributes to acid rain is nitrogen dioxide. This is formed by a reaction between nitrogen and oxygen in the air whenever a high enough temperature is reached.

$$N_2(g) + 2O_2(g) \rightarrow 2NO_2(g)$$

Nitrogen dioxide is a reddish-brown, poisonous gas and, like sulphur dioxide, it has a sharp, choking smell. As we have seen already (in Section 2), nitrogen dioxide is formed in petrol engines in the region of the spark caused by the sparkplugs, but it is also formed in industrial furnaces.

Figure 13.2 ▲
Over 5 million tonnes of coal is burned at Longannet power station every year.

For example, the furnaces at a coal-fired power station reach a sufficiently high temperature to allow nitrogen dioxide to form.

When nitrogen dioxide dissolves in rain water it produces nitric acid which, like sulphuric acid, is a strong acid.

What damage does acid rain cause?

Buildings

Acid rain has damaging effects on buildings made from carbonate rocks, such as limestone and marble, which are both made of calcium carbonate. Calcium carbonate is also a cementing substance in sandstone, so as acid rain wears it away, the sandstone crumbles.

In Egypt, the 4,600 year old statue of the Sphinx (see Figure 13.3) is made mostly of limestone. Old photographs show that many of the features of the Sphinx have been eroded away in recent years.

In India, much of the Taj Mahal is made of marble which is also under attack from acid pollution. It is only 48 km away from an oil refinery which is allowed to release large quantities of sulphur dioxide into the air. In addition, the Taj Mahal is close to busy roads with thousands of vehicles using them daily. Like the Sphinx, the Taj Mahal is suffering from the effects of acid erosion.

Many of our fine buildings are showing signs of attack by acid rain. This is because they were often built using sandstone in which the calcium carbonate cementing material can be attacked by the sulphuric acid (and nitric acid) in acid rain.

$$CaCO_3(s) + H_2SO_4(aq) \rightarrow CaSO_4(aq) + H_2O(l) + CO_2(g)$$

Figure 13.3 ▲

Many of the features of the Sphinx have been eroded by acid rain.

The calcium sulphate ($CaSO_4$) can sometimes be seen as a white coating on the surface of sandstone, or as a white stalactite under a sandstone arch or bridge. This happens when calcium sulphate solution reaches the surface of the stone and water evaporates to leave the solid calcium sulphate.

Iron and steel structures

Iiron and steel (steel is mainly iron) rust when exposed to air, water and electrolyte (see Chapter 14). Carbonic acid can act as the electrolyte to allow rusting to take place, but since sulphuric acid and nitric acid are stronger, they provide more ions and are better electrolytes. As a result, when acid rain comes into contact with exposed iron or steel, rusting takes place even faster. Both acids can attack the iron directly forming a salt and hydrogen gas.

$$Fe(s) + H_2SO_4(aq) \rightarrow FeSO_4(aq) + H_2(g)$$

Living things

Many lochs and lakes in Scotland, Norway, Sweden and elsewhere are now fishless due to acid rain. Some have pH values as low as 4, but the acidity does not kill the fish directly. Aluminium ions are released by the low acidity from the soil surrounding the lochs and lakes. Aluminium ions poison the fish by causing their gills to become covered with mucous. This prevents them from absorbing dissolved oxygen in the water.

Acid rain can harm trees in two main ways. First, the acidity damages leaves and pine needles making them turn yellow and drop off. Second, important nutrient ions, such as calcium and magnesium, are leached out of the soil. It also seems likely that the trees absorb poisonous aluminium ions from the soil, which have been released by the acidity. A huge number of trees around the world have been damaged by acid rain (see Figure 13.4).

Figure 13.4 ▲
Trees damaged by acid rain.

Figure 13.5 ▲
Why is lime being added to this Swedish lake?

The pH of soil can be lowered so much by acid rain that many crops cannot be grown. Fortunately, powdered lime (calcium oxide) or limestone (calcium carbonate) can be added to neutralise the acidity and increase soil pH. Chalk, which is readily available in Southern England can also be used for the same purpose because, like limestone and marble, it is another form of calcium carbonate. The same materials can be added successfully to areas of water so that fish can live and spawn successfully (see Figure 13.5).

Solving the problem of acid rain

Reference has already been made on pages 24 and 79 to the use of catalytic converters in the exhaust systems of petrol engines. These help to reduce the problem of nitrogen dioxide. However, the largest single source of sulphur dioxide is coal-fired power stations. Since it has not proved to be possible to remove sulphur from coal before it is burned, a process has been developed for removing sulphur dioxide from the flue gases. The huge Drax power station in Yorkshire has this 'flue gas desulphurisation' process in operation.

Air is mixed with the flue gases and the mixture is passed through a mixture of powdered limestone and water. The sulphur dioxide reacts with water to form sulphurous acid.

$$H_2O(l) + SO_2(g) \rightarrow H_2SO_3(aq)$$

The sulphurous acid reacts with the limestone (calcium carbonate) to give calcium sulphite, water and carbon dioxide.

$$H_2SO_3(aq) + CaCO_3(s) \rightarrow CaSO_3(s) + H_2O(l) + CO_2(g)$$

Finally, the calcium sulphite reacts with oxygen to give calcium sulphate.

$$2CaSO_3(s) + O_2(g) \rightarrow 2CaSO_4(s)$$

Calcium sulphate is a useful product since it can be used in the production of plasterboard for use in house construction.

Questions

5 **(a)** Which gas causes rain water to be slightly acidic naturally?
(b) Name the two gases mainly responsible for acid rain.

6 **(a)** Name the main acid in acid rain.
(b) Why are limestone and marble attacked by this acid in acid rain?

7 How is the emission of acidic gases from petrol engines and coal-fired power stations being reduced?

Figure 13.6 ▲
Fertiliser bags show the percentage of each of three important nutrients present (NPK).

Making fertilisers by neutralisation

You may have come across ammonia solution, not only as a smelly alkaline solution in the laboratory, but also as a household cleaning agent. Ammonia solution is even more versatile – farmers use it as a fertiliser because it not only neutralises soil acidity, raising pH, but also acts as a rich source of nitrogen, which plants need in order to make plant proteins. However, if there is no need to raise the pH of the soil, ammonia in the form of ammonium salts can be used as fertilisers. The ammonium salts used are water-soluble, not smelly and are available as easily distributed granules, so they are much easier to deal with than ammonia solution. There are three widely used ammonium fertilisers – ammonium nitrate, ammonium sulphate and ammonium phosphate.

Ammonia reacts with nitric acid to give ammonium nitrate.

$$NH_3 + HNO_3 \rightarrow NH_4NO_3$$

Ammonia reacts with sulphuric acid to give ammonium sulphate.

$$2NH_3 + H_2SO_4 \rightarrow (NH_4)_2SO_4$$

Ammonia reacts with phosphoric acid to give ammonium phosphate which contains another important plant nutrient, phosphorus.

$$3NH_3 + H_3PO_4 \rightarrow (NH_4)_3PO_4$$

Note: These are simplified formula equations. In each of them hydroxide ions from the alkali ammonia solution react with hydrogen ions from the acid to produce water in addition to the reaction shown. As a result it is true to say that 'acid + alkali → salt + water'.

Questions

8 (a) Why might ammonia solution be preferred to ammonium sulphate for use as a fertiliser?

(b) Why might ammonium sulphate be preferred to ammonia solution for use as a fertiliser.

9 Write a simplified formula equation for the reaction of ammonia and hydrochloric acid to give ammonium chloride.

10 Potassium nitrate is a suitable form for the plant nutrient potassium as well as nitrogen. Write a formula equation to show its formation by means of a reaction between an acid and an alkali.

Making soluble salts by neutralisation

Using alkalis (soluble bases)

Solutions of acids and alkalis are colourless, so although we know that they will react to form a salt and water, how will we know when neutralisation is exact? The answer could be to use some pH paper, but we must not dip this into the solution or the indicators present will 'bleed' into the liquid and result in the final salt being coloured and not white, as it should be. Read through Figure 13.7 to see how to make crystals of a soluble salt by neutralisation using an alkali.

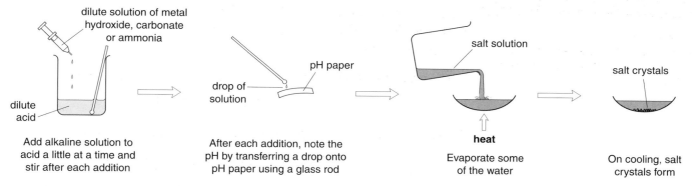

Figure 13.7 ▲
Making soluble salts using an acid and an alkali (soluble base).

This technique could be used, for example, to make samples of the ammonium salts that are used as fertilisers.

Question

11 Using the method described in Figure 13.7, it is possible to make crystals of sodium sulphate by reaction between sodium hydroxide solution and dilute sulphuric acid. Write a balanced formula equation for this reaction.

Using insoluble bases or metals

Insoluble bases and suitable metals, such as magnesium and zinc, can also be used to make soluble salts. There is, however, no need to test drops of solution at intervals with pH paper, because excess solid that is left unreacted can easily be filtered off and removed. It is easy to see when the reaction is over if a metal carbonate or metal is being used. This is because they give off gases when reacting with acids, so if unreacted solid is present and no gas is being given off, then it can be assumed that the reaction is over. Figure 13.8 shows how to make crystals of a soluble salt by neutralisation using a metal or insoluble base.

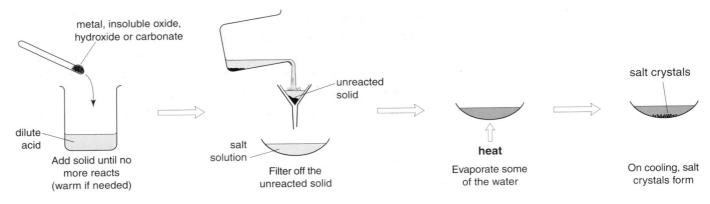

Figure 13.8 ▲

Making soluble salts using an acid and insoluble base or metal.

Prescribed Practical Activity

Safety note

Acid mists are formed.

Hydrogen is flammable.

Preparation of a salt

- The aim of this experiment is to prepare a pure sample of a soluble salt such as magnesium sulphate.
- Magnesium metal is added to a given volume of a dilute acid such as sulphuric acid until there is no further evolution of hydrogen gas.
- The excess metal is filtered off and some of the water is evaporated from the salt solution obtained as filtrate.
- The hot solution is set aside to crystallise.
- The same method can also be used with an insoluble base such as magnesium carbonate. In this case carbon dioxide is given off.
- Copper(II) carbonate can be reacted successfully to give copper(II) sulphate, but copper metal does *not* react with a dilute acid such as dilute sulphuric acid.

Question

12 Write balanced formula equations showing state symbols for the reactions between dilute sulphuric acid and
  (a) magnesium,
  (b) magnesium carbonate
  (c) copper(II) carbonate.

Making insoluble salts by precipitation

An insoluble salt is one where less than 1 g of the salt can dissolve in 1 litre of water at 20°C. In data booklets, tables of solubility may show this property by means of the letter 'i', indicating insolubility. We can make insoluble salts by using a type of reaction called a **precipitation reaction**.

A precipitation reaction is one in which two solutions react to form an insoluble product called a **precipitate**. Silver(I) iodide, for example, is insoluble and can be made by the following method:

A soluble silver(I) salt is dissolved in water, and a soluble iodide salt is also dissolved in water. When these two solutions are mixed, a precipitate of silver(I) iodide forms (see Figure 13.9 for this and other steps in the procedure). In order to obtain a pure, dry sample of this salt, it is filtered off, washed with deionised or distilled water and allowed to dry in air, or in an oven. The balanced formula equation for the reaction taking place is:

$$AgNO_3(aq) + NaI(aq) \rightarrow AgI(s) + NaNO_3(aq)$$

Ionic equation:

$$Ag^+(aq) + NO_3^-(aq) + Na^+(aq) + I^-(aq) \rightarrow$$
$$Ag^+I^-(s) + Na^+(aq) + NO_3^-(aq)$$

Removing spectator ions gives:

$$Ag^+(aq) + I^-(aq) \rightarrow Ag^+I^-(s)$$

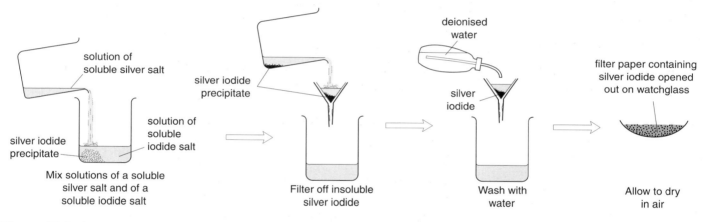

Figure 13.9 ▲
Making the insoluble salt, silver(I) iodide, using precipitation.

Questions

13 Write simple formulae equations, with state symbols, for the reaction between:
(a) potassium chloride solution and silver(I) nitrate solution,
(b) barium nitrate solution and sodium sulphate solution.

14 Write ionic equations, after removal of spectator ions, showing state symbols, for the reactions between:
(a) copper(II) sulphate solution and lithium carbonate solution,
(b) lead(II) nitrate solution and sodium iodide solution.

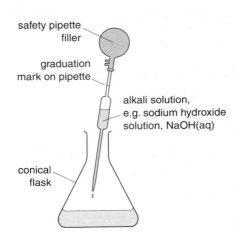

Figure 13.10 ▲
Pipette the alkali.

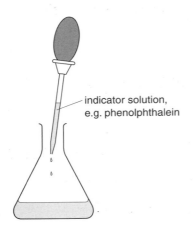

Figure 13.11 ▲
Add indicator.

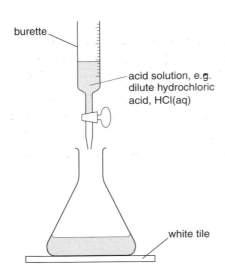

Figure 13.12 ▲
Run in acid from burette.

Volumetric titrations

How can you work out the concentration of a solution of an acid or an alkali if you do not know the mass of solute that was dissolved? The answer is to carry out a **volumetric titration**. This is an experiment in which the volumes of reacting solutions are measured accurately. Using a volumetric titration, which is often just referred to as a titration, it is possible to calculate the concentration of an acid if the concentration of an alkali is known, and vice versa.

How to carry out a titration

Two special pieces of equipment are needed for the measurement of volume. A **pipette** is used to measure out an accurate volume of alkali solution (see Figure 13.10). The liquid is drawn into the pipette using a safety filler. A **burette** is used to measure the volume of acid solution that is required to neutralise the alkali (see Figure 13.12).

Indicators are compounds whose colour is dependent on pH. Phenolphthalein is a suitable indicator for use in a titration involving a strong acid and a strong alkali. Phenolphthalein is pink in a strongly alkaline solution, but loses its colour completely if even one drop of acid is in excess.

To perform the titration, a known volume of alkali solution is pipetted into a conical flask, 2 or 3 drops of indicator are added and a trial titration is carried out. Acid is run in from the burette $1 \, cm^3$ at a time, with the flask being swirled after each addition. Eventually the indicator changes colour, the burette reading is noted and the flask is washed out. The trial titration establishes roughly what volume of alkali is needed to neutralise the acid.

The titration procedure is then repeated, but acid can now be run in quickly to within $1 \, cm^3$ of the trial value and then added drop-wise, with swirling after each addition. The **end point** of the titration, when the reaction is complete, can now be established much more accurately as shown by the colour change in the indicator. It is normal practice to continue titrating until you have two concordant results (ones that agree within $0.2 \, cm^3$). The average of the concordant results is calculated, with the trial result and any other high or low results being ignored.

Using the average value for the volume of acid required to neutralise the alkali, it is possible to calculate the concentration of one solution if that of the other solution is known.

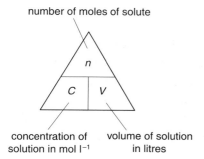

number of moles of solute

n

C V

concentration of
solution in mol l^{-1}

volume of solution
in litres

Question

15 (a) What is meant by 'the end point' of a titration?

(b) Three titration values were 19.0 cm^3, 18.2 cm^3 and 18.4 cm^3. What average value should be recorded?

Calculations based on titration results

The relationship between concentration of solution (C), number of moles of solute (n) and volume of solution (V) can be used in calculations based on volumetric titration results.

$$n = C \times V \qquad C = \frac{n}{V} \qquad V = \frac{n}{C}$$

Example 1

20 cm^3 of sodium hydroxide solution was neutralised by 15 cm^3 of 0.1 mol l^{-1} sulphuric acid. Calculate the concentration of the sodium hydroxide solution.

Balanced equation for the reaction:

$$H_2SO_4 + 2NaOH \rightarrow Na_2SO_4 + 2H_2O$$
$$\text{1 mol} \qquad \text{2 mol}$$

Number of moles of acid reacting, $n_{H_2SO_4} = C \times V$
$$= 0.1 \times 0.015 = 0.0015$$

From the equation: 1 mol H_2SO_4 reacts with 2 mol NaOH.

Thus: 0.0015 mol H_2SO_4 reacts with 0.0030 mol NaOH

Concentration of NaOH, $C_{NaOH} = \frac{n}{V}_{NaOH} = \frac{0.0030}{0.02} = 0.15$ mol l^{-1}

Alternative method

When carrying out titrations, we usually know the volume of acid (V_A) and the volume of alkali or soluble base (V_B). We also know either the concentration of the acid (C_A) or the concentration of the alkali (C_B). From the balanced equation for the reaction, we can find the number of moles of acid (a) and the number of moles of alkali (b) that react together. Using the relationship $n = C \times V$, we can write the ratio of moles of acid to moles of alkali required for neutralisation as:

$$\frac{C_A \times V_A}{C_B \times V_B} = \frac{a}{b}$$

From this relationship we can calculate any of the six quantities if the other five are known.

Note: Because the relationship is a ratio, there is no need to convert the volumes into litres. Any units of volume may be used for V_A and V_B but they must be the same units in each case.

Example 2

Use the alternative method for the calculation given in the previous example.

Balanced equation for the reaction:

$$H_2SO_4 + 2NaOH \rightarrow Na_2SO_4 + 2H_2O$$

$$\begin{array}{cc} 1\,\text{mol} & 2\,\text{mol} \\ a = 1 & b = 2 \end{array}$$

Rearranging $\dfrac{C_A \times V_A}{C_B \times V_B} = \dfrac{a}{b}$ gives: $C_B = \dfrac{C_A \times V_A \times b}{V_B \times a}$

$$= \frac{0.1 \times 15 \times 2}{20 \times 1} = 0.15 \,\text{mol}\,1^{-1}$$

Questions

16 $25\,\text{cm}^3$ of $0.2\,\text{mol}\,1^{-1}$ sodium hydroxide solution was neutralised by $16\,\text{cm}^3$ of hydrochloric acid. Calculate the concentration of the hydrochloric acid.

17 $10\,\text{cm}^3$ of potassium hydroxide solution was neutralised by $12\,\text{cm}^3$ of $0.1\,\text{mol}\,1^{-1}$ nitric acid. Calculate the concentration of the potassium hydroxide solution.

18 $20\,\text{cm}^3$ of $0.5\,\text{mol}\,1^{-1}$ lithium hydroxide solution was neutralised by $30\,\text{cm}^3$ of sulphuric acid. Calculate the concentration of the sulphuric acid.

19 $15\,\text{cm}^3$ of ammonia solution was neutralised by $12.5\,\text{cm}^3$ of $0.15\,\text{mol}\,1^{-1}$ phosphoric acid. Calculate the concentration of the ammonia solution. The balanced formula equation for the reaction is as follows:

$$3NH_3 + H_3PO_4 \rightarrow (NH_4)_3PO_4$$

In questions 1–4 choose the correct word **from the following list** to complete the sentence.

carbon, burette, non-metal, alkali, sulphur, pipette, metal, acid

1 The neutralisation of a solution of an _____ moves the pH down towards 7.

2 A salt is a compound in which the hydrogen ion of an acid has been replaced by a _____ ion or an ammonium ion.

3 A _____ is used to measure an alkali solution into a flask before titration against an acid solution.

4 The gas that is largely responsible for acid rain is _____ dioxide.

5 When a dilute acid reacts with a carbonate, hydrogen ions and carbonate ions react to produce:
    A a salt and carbon dioxide
    B hydrogen and carbon dioxide
    C a salt and water
    D water and carbon dioxide

6 Which of the following equations is balanced?
    A $CuO + HNO_3 \rightarrow Cu(NO_3)_2 + H_2O$
    B $2CuO + HNO_3 \rightarrow 2Cu(NO_3)_2 + H_2O$
    C $CuO + 2HNO_3 \rightarrow Cu(NO_3)_2 + H_2O$
    D $2CuO + 2HNO_3 \rightarrow Cu(NO_3)_2 + 2H_2O$

7 Which of the following pairs of solutions would you expect to react forming a precipitate?
    A zinc(II) iodide and ammonium phosphate
    B aluminium nitrate and lithium chloride
    C iron(II) sulphate and nickel(II) bromide
    D sodium carbonate and potassium sulphate

8 $20\,cm^3$ of $0.2\,mol\,l^{-1}$ nitric acid is exactly neutralised by $0.4\,mol\,l^{-1}$ potassium hydroxide. The volume of the potassium hydroxide solution is:
    A $80\ cm^3$
    B $40\ cm^3$
    C $20\ cm^3$
    D $10\ cm^3$

9 The balanced equation for the reaction between phosphoric acid and sodium hydroxide is:

$$3NaOH + H_3PO_4 \rightarrow Na_3PO_4 + 3H_2O$$

The volume of $0.1\,mol\,l^{-1}$ NaOH that is required to neutralise $40\,cm^3$ of $0.05\,mol\,l^{-1}$ phosphoric acid is:
    A $60\,cm^3$
    B $40\,cm^3$
    C $30\,cm^3$
    D $20\,cm^3$

10 (a) What name is given to bases which dissolve in water?
    (b) Write a balanced formula equation for the reaction of sodium carbonate with dilute hydrochloric acid.

11 (a) Hydrogen gas is given off in reactions between certain metals and most acids. Describe the chemical test for hydrogen.
    (b) Give a balanced formula equation for the reaction between magnesium and dilute hydrochloric acid.

12 (a) Describe *two* of the damaging effects of acid rain.
    (b) Which gas dissolves to produce the nitric acid that can be present in acid rain?

13 (a) A fertiliser contains the nutrients nitrogen (20%), phosphorus (15%) and potassium (5%). Present this information in the form of a bar graph.
    (b) Why is ammonium phosphate a faster acting fertiliser than calcium phosphate? (You may wish to refer to a data booklet.)

14 (a) Name the precipitate that is formed when a solution of tin(II) sulphate is added to a solution of potassium carbonate.
    (b) Name two salt solutions that could be mixed in order to produce a precipitate of lead(II) iodide.

15 $25\,cm^3$ of $0.1\,mol\,l^{-1}$ potassium hydroxide is required to neutralise $17.5\,cm^3$ of hydrochloric acid. Calculate the concentration of the hydrochloric acid.

Show your working clearly.

14 Metals

The electrochemical series

Batteries and cells

How many things can you think of that are 'battery operated'? Computer games, mobile phones, remote controls, torches, toys, radios, miniature televisions, watches and clocks are only a few (see Figure 14.1). Batteries are **chemical cells**, which convert chemical energy into electrical energy. A **battery** consists of two or more chemical cells joined together. We tend to use the words 'battery' and 'cell' interchangeably.

Figure 14.1 ▲
All of these are battery operated.

Chemical cells are quite simple in essence. Within any one there is a substance that gives up electrons and another that accepts them. Electrons flow from one substance to the other through metal wires and other parts of the device to be operated, such as a torch bulb (see Figure 14.2). In order that a current can flow, a special conductor that does not allow electrons to move through it must complete the circuit. The special conductor used is an **electrolyte**, since this conducts by movement of ions instead of electrons.

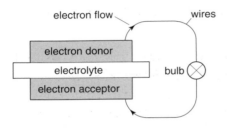

Figure 14.2 ▲
A chemical cell – how it works.

Although their shapes may be different from that in Figure 14.2, all chemical cells work because of electron loss and electron gain. In rechargeable batteries the direction of electron flow can be reversed, reforming the original chemicals and thus recharging the cell. Some chemical cells can be recharged thousands of times.

Mains electricity and batteries compared

The voltage of an electrical supply is a measure of the energy that the electrons possess. Mains voltage is 240 volts and is so dangerous that it can kill. Batteries, on the other hand, are much safer, with typical voltages seldom being greater than 12 volts. Mains electricity is very much cheaper than that obtained from ordinary batteries, which cannot be recharged, but that is the penalty which must be paid for portability. Think what life would be like, for some people at least, if they were not able to listen to their favourite music no matter where they are.

The burning of fossil fuels at power stations is harmful to the environment, but the extraction of metals for the production of batteries also causes harm. Some of the metals used for battery production, such as zinc, lead and nickel, are becoming scarcer. Batteries are greater users of these finite resources than power stations making mains electricity. Some metals, like lead and cadmium are poisonous and batteries containing them should be returned for recycling to avoid toxic material getting into the environment.

Figure 14.3 ▲

Two different metals dipping into electrolytes provide electricity for this clock.

Metal X	Voltage
magnesium	+2.1
zinc	+1.0
iron	+0.7
tin	+0.4
lead	+0.3
copper	0.0
silver	−0.4
gold	−0.6

Table 14.1 ▲

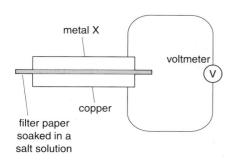

Figure 14.4 ▲

A simple cell for measuring the voltage produced by two metals.

Metals and the electrochemical series (ECS)

Have you ever seen a clock like the one in Figure 14.3? The makers call it a 'lemon clock', but it works using other fruit and vegetables as well. Two strips of metal, one copper and the other zinc, are pushed into a vegetable such as a potato, or a fruit such as a lemon. A digital clock is then attached to the metal strips. A chemical cell has been created with electrons flowing from the zinc strip to the copper strip through the wires and clock. The fruit or vegetable used provides electrolytes that are needed to complete the circuit. This is because they contain ions, with enough water present to allow the ions to move. If a voltmeter is used in place of the clock a reading of about 1 volt is obtained.

In place of the fruit or vegetable we can use a piece of filter paper soaked in a salt solution, such as sodium chloride solution. The two metals can then be placed on either side of this, in contact with it (see Figure 14.4). If copper is retained as a standard and other metal elements (metal X in Figure 14.4) are used in place of zinc, we find that different pairs of metals give different voltages. A typical table of results is shown in Table 14.1.

The 'league table of metals' sorted according to the voltages produced by different pairs of metals is called the **electrochemical series**, or **ECS**. The higher in the ECS a metal is, the more readily it loses electrons. When two different metals are used in the cell shown in Figure 14.4, electrons flow *from the metal higher* in the ECS to the one lower down through the wires and meter. We can also see from the results that the further apart two metals are in the ECS the larger is the cell voltage.

It is possible to include all metal elements in an electrochemical series, but usually a simpler one is used like the one shown in Table 14.2. This includes some common metals that were not given in Table 14.1.

Questions

1   In a Zn/Cu cell, electrons flow from the zinc to the copper through the wires and meter. Give the direction of electron flow in each of the following cells.

(a) Mg/Fe; (b) Ag/Sn; (c) Ni/Zn; (d) Al/Pb.

2   For each of the following state which cell would give the larger voltage.

(a) Ni/Pb and Fe/Cu
(b) Fe/Ag and Sn/Ag
(c) Al/Hg and Mg/Au
(d) Zn/Pb and Al/Hg

Name	Symbol
lithium	Li
potassium	K
calcium	Ca
sodium	Na
magnesium	Mg
aluminium	Al
zinc	Zn
iron	Fe
nickel	Ni
tin	Sn
lead	Pb
copper	Cu
silver	Ag
mercury	Hg
gold	Au

Table 14.2 ▲
An ECS of metals.

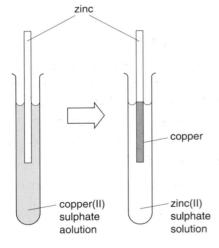

Figure 14.5 ▲
Zinc displaces copper from copper(II) sulphate solution.

Displacement and the ECS

Here is a chemical puzzle. Why is it that if a piece of silvery grey zinc is placed in a solution of blue copper(II) sulphate, the zinc is eaten away and becomes covered in brown copper? At the same time, the blue colour of the solution slowly fades (see Figure 14.5). What happens is that zinc atoms each lose two electrons to form zinc(II) ions, which are colourless and go into solution. The blue copper(II) ions each gain two electrons, that were given up by the zinc atoms, and form layers of brown copper atoms on the zinc. We say that zinc *displaces* the copper and we can show what happens using ion-electron equations.

$$Zn(s) \rightarrow Zn^{2+}(aq) + 2e^-$$
zinc atoms   zinc(II) ions

$$Cu^{2+}(aq) + 2e^- \rightarrow Cu(s)$$
copper(II) ions   copper atoms

Adding the two ion-electron equations together and omitting electrons gives the following overall equation for the reaction.

$$Zn(s) + Cu^{2+}(aq) \rightarrow Zn^{2+}(aq) + Cu(s)$$
zinc atoms   copper(II) ions   zinc(II) ions   copper atoms
(silvery grey)   (blue)   (colourless)   (brown)

If the experiment is changed and copper metal is added to a solution containing zinc(II) ions, you find there is no reaction. However, if magnesium is added to a solution containing zinc(II) ions, then zinc is displaced.

A **displacement reaction** involves the formation of a metal from a solution containing its ions by reaction with a metal *higher* in the ECS. The **displacement rule** is that a metal *higher* in the ECS will displace one that is lower down from a solution containing the ions of the lower metal. The general word equation for any displacement reaction is as follows:

atoms of   +   ions of   →   ions of   +   atoms of
higher metal   lower metal   higher metal   lower metal

Thus, adding magnesium to a solution containing zinc(II) ions produces magnesium ions and zinc atoms, which can be written as a balanced formula equation:

$$Mg(s) + Zn^{2+}(aq) \rightarrow Mg^{2+}(aq) + Zn(s)$$

Displacement and the position of hydrogen in the ECS

The reaction between metals and acids can be used to establish the position of hydrogen in the ECS. Figure 14.6 shows the results of adding dilute hydrochloric acid to the four metals zinc, iron, lead and copper.

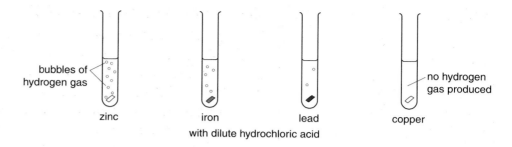

bubbles of hydrogen gas

no hydrogen gas produced

zinc   iron   lead   copper

with dilute hydrochloric acid

Figure 14.6 ▲
Reactions of metals with dilute hydrochloric acid.

Name	Symbol
lithium	Li
.	.
.	.
lead	Pb
hydrogen	**H$_2$**
copper	Cu
.	.
.	.
gold	Au

Table 14.3 ▲
The position of hydrogen in the ECS.

Where a reaction takes place, producing hydrogen gas, the equations closely resemble displacement reactions in which metals are displaced, for example:

$$Fe(s) + 2H^+(aq) \rightarrow Fe^{2+}(aq) + H_2(g)$$

Lead and all metals above lead in the ECS displace hydrogen from acids, but copper and all metals below copper do not displace hydrogen from acids. As a result hydrogen is placed in the ECS between lead and copper (see Table 14.3).

Oxidation, reduction and redox reactions

So far in this chapter we have been concerned with three main types of reaction – those in which electrons are lost, those in which they are gained, and those in which electrons are lost by one substance and gained by another. An **oxidation reaction** is one in which electrons are *lost*. An example of this is the loss of electrons by zinc atoms when displacing copper from a solution containing copper(II) ions:

$$Zn(s) \rightarrow Zn^{2+}(aq) + 2e^-$$

Because they lose electrons, the zinc atoms are said to have been **oxidised**.

A **reduction reaction** is one in which a substance *gains* electrons. The substance is then said to have been **reduced**. In the example given above, there is a gain of electrons by the copper(II) ions, which are, therefore, reduced:

$$Cu^{2+}(aq) + 2e^- \rightarrow Cu(s)$$

In a **redox reaction**, both oxidation and reduction take place. Electrons are lost by one substance and gained by another. The reaction between zinc atoms and copper(II) ions is therefore an example of a redox reaction because electrons are lost by zinc atoms and gained by copper(II) ions. The overall redox reaction is

$$Zn(s) + Cu^{2+}(aq) \rightarrow Zn^{2+}(aq) + Cu(s)$$

Oxidation
Is
Loss (of electrons)

Reduction
Is
Gain (of electrons

Figure 14.7 ▲
'OIL RIG' – a useful mnemonic.

It is worth noting that all displacement reactions are redox reactions, involving the oxidation of one substance and the reduction of another. The displacement of hydrogen from acids by metals is therefore also described as redox. For example when magnesium reacts with dilute sulphuric acid the following reaction takes place:

$$Mg(s) + 2H^+(aq) \rightarrow Mg^{2+}(aq) + H_2(g) \text{ (redox)}$$

Each magnesium atom loses two electrons:

$$Mg(s) \rightarrow Mg^{2+}(aq) + 2e^- \text{ (oxidation)}$$

Two hydrogen ions accept the two electrons and form a hydrogen molecule:

$$2H^+(aq) + 2e^- \rightarrow H_2(g) \text{ (reduction)}$$

A useful way of remembering the definitions of oxidation and reduction is to use 'OIL RIG' (see Figure 14.7).

More about oxidation and reduction

Later in this chapter we look at the extraction of metals. Most occur in compounds and the problem is how to form the metal element from the compound. In reality, it is a case of 'persuading' positive metal ions to accept electrons and form metal atoms. When a metal is extracted from a compound, metal ions are *reduced*. Because this is the most important reaction taking place, any reaction of a *compound* to form a *metal* is considered to be an example of **reduction**. By similar reasoning, any reaction of a *metal* to form a *compound* is considered to be an example of **oxidation**. This is because the most important reaction taking place is the loss of electrons by metal atoms to form positive metal ions in the compound.

More complex cells

Electricity can be produced in a cell by connecting beakers which contain two different metals dipping into solutions containing their respective ions as in Figure 14.8.

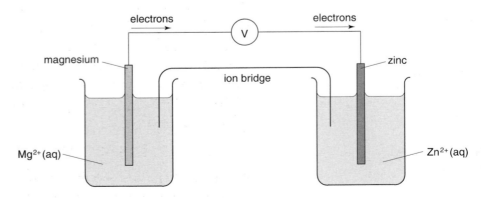

Figure 14.8 ▲
A magnesium/zinc cell.

Questions

3 **(a)** State the 'displacement rule'.
   (b) In each of the following examples, will displacement take place?
   i)   magnesium + zinc(II) sulphate solution
   ii) tin + iron(II) sulphate solution
   iii) lead + sodium nitrate solution
   iv) nickel + copper(II) chloride solution

   (c) Where displacement takes place in part b)
   i)   Write formula equation(s) for the redox reactions taking place, omitting spectator ions.
   ii) Write ion-electron equations for the oxidation and reduction reactions involved.

4 For the reaction between zinc and dilute sulphuric acid

   (a) Write a word equation for the reaction.
   (b) Name the spectator ion in the reaction.
   (c) Write a formula equation for the redox reaction omitting the spectator ion.
   (d) Give the relevant ion-electron equations associated with the reaction, adding 'oxidation' and 'reduction' as appropriate.

5 For each of the following reactions, decide whether they should be labelled 'oxidation', 'reduction', 'redox' or 'other'.

   (a) $Mg(s) + Sn^{2+}(aq) \rightarrow Mg^{2+}(aq) + Sn(s)$
   (b) $2H^+(aq) + CO_3^{2-}(aq) \rightarrow H_2O(l) + CO_2(g)$
   (c) $2e^- + Br_2(aq) \rightarrow 2Br^-(aq)$
   (d) $SO_3^{2-}(aq) + H_2O(l) \rightarrow SO_4^{2-}(aq) + 2H^+(aq) + 2e^-$

The metals are connected by wires and a meter, while the solutions are connected by means of an **ion bridge**. This can simply be a piece of filter paper soaked in a salt solution and its purpose is the same as in the simple cell in Figure 14.4, namely, to complete the circuit. Electrons flow from the metal higher in the ECS to the one lower down through the wires and meter. In the cell illustrated the electrons therefore flow as shown from the magnesium to the zinc. The metal higher in the ECS, magnesium, loses electrons and passes into solution as ions:

$$Mg(s) \rightarrow Mg^{2+}(aq) + 2e^- \text{ (oxidation)}$$

The ions of the metal lower in the ECS, which in this case is zinc, gain electrons and form metal atoms:

$$Zn^{2+}(aq) + 2e^- \rightarrow Zn(s) \text{ (reduction)}$$

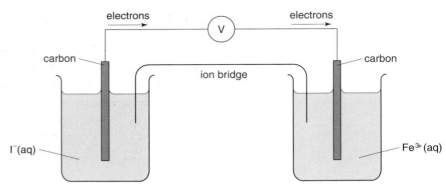

Figure 14.9 ▲
An iodide ion/iron(III) ion cell.

Electricity can be produced in cells which have only one metal or, in some cases, no metals at all. In such cases carbon rods are used in place of metals to make contact with chemicals in solution that can either lose or gain electrons. In the cell shown in Figure 14.9 a solution containing iodide ions, I^-(aq), is present in one beaker and a solution containing iron(III) ions, Fe^{3+}(aq), is present in the other. Iodide ions act as electron donors, giving up electrons. These flow through the wires and meter to the iron(III) ions, which accept them. The reactions taking place at the two carbon rods are as follows:

$$2I^-(aq) \rightarrow I_2(aq) + 2e^- \text{ (oxidation)}$$

$$2e^- + 2Fe^{3+}(aq) \rightarrow 2Fe^{2+}(aq) \text{ (reduction)}$$

The formation of iodine can be shown by the addition of starch solution, with which it gives a dark blue colour. Later in this chapter on page 201 you will learn that **ferroxyl indicator** can be used to show the presence of iron(II) ions in solution, Fe^{2+}(aq), with which it also forms a blue colour.

Iodide ions and iron(III) ions can be included in the ECS along with other species such as the molecules of non-metals. You will be able to find the halogen molecules, for example, included in the ECS that is provided in most data booklets. You will also notice that the information given in the booklets enables you to work out the ion-electron equations for reactions taking place in cells.

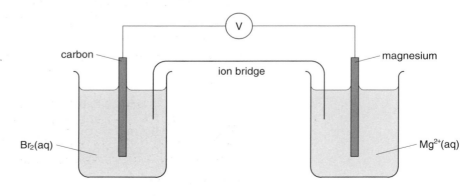

Figure 14.10 ▲
A magnesium/bromine cell.

Prescribed Practical Activity

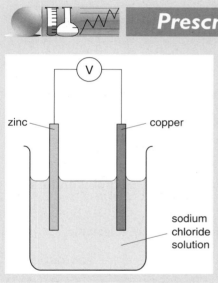

Factors which affect voltage

- The aim of this experiment is to investigate a factor which might affect the size of the voltage produced by a simple cell.
- One example of such a factor is to find out if different pairs of metals produce different voltages.
- A cell like the one shown can be used.
- The zinc can be replaced by another metal, such as iron or magnesium, and the voltage noted again.
- Alternatively, zinc and copper can be retained and the electrolyte can be changed, the sodium chloride solution being replaced by hydrochloric acid or sodium hydroxide solution, for example.

Redox equations from ion-electron equations

As we saw on page 186, ion-electron equations for the oxidation and reduction reactions taking place can be combined to give the redox equation for the overall reaction. In order to do this the number of electrons lost and gained must be the same.

Example 1
If the ion-electron equations contain equal numbers of electrons, they are simply added together to give the overall redox equation with the electrons omitted. As an example of this we can look again at the displacement reaction that takes place between zinc and copper(II) ions.

Ion-electron equations:

I $Zn(s) \rightarrow Zn^{2+}(aq) + 2e^-$ (oxidation)

II $2e^- + Cu^{2+}(aq) \rightarrow Cu(s)$ (reduction)

Adding I and II (and omitting electrons):

$Zn(s) + Cu^{2+}(aq) \rightarrow Zn^{2+}(aq) + Cu(s)$ (redox)

Example 2

If the ion-electron equations do not contain equal numbers of electrons, then the electrons are balanced by mutiplying one or both of the ion-electron equations. Consider, for example, the displacement reaction involving copper and silver(I) ions:

Ion-electron equations:

I $Cu(s) \rightarrow Cu^{2+}(aq) + 2e^-$ (oxidation)

II $e^- + Ag^+(aq) \rightarrow Ag(s)$ (reduction)

Multiplying equation II by 2:

$2e^- + 2Ag^+(aq) \rightarrow 2Ag(s)$

Adding this to equation I (and omitting electrons):

$Cu(s) + 2Ag^+(aq) \rightarrow Cu^{2+}(aq) + 2Ag(s)$ (redox)

Example 3

The same process can be applied to more complex examples, such as the reaction between hydrogen peroxide and permanganate ions in acid solution.

Ion-electron equations (state symbols omitted for clarity):

I $H_2O_2 \rightarrow O_2 + 2H^+ + 2e^-$ (oxidation)

II $5e^- + MnO_4^- + 8H^+ \rightarrow Mn^{2+} + 4H_2O$ (reduction)

Multiply equation I by 5 and equation II by 2 to give:

I $5H_2O_2 \rightarrow 5O_2 + 10H^+ + 10e^-$

II $10e^- + 2MnO_4^- + 16H^+ \rightarrow 2Mn^{2+} + 8H_2O$

Adding these equations (and omitting electrons):

$5H_2O_2 + 2MnO_4^- + 16H^+ \rightarrow 5O_2 + 10H^+ + 2Mn^{2+} + 8H_2O$

In this equation $10H^+$ ions may also be removed from both sides, giving the final overall equation:

$5H_2O_2 + 2MnO_4^- + 6H^+ \rightarrow 5O_2 + 2Mn^{2+} + 8H_2O$ (redox)

Question

8 Write overall redox equations for each pair of ion-electron equations.

(a) $Mg(s) \rightarrow Mg^{2+}(aq) + 2e^-$

$Pb^{2+}(aq) + 2e^- \rightarrow Pb(s)$

(b) $Fe^{2+}(aq) \rightarrow Fe^{3+}(aq) + e^-$

$Cl_2(aq) + 2e^- \rightarrow 2Cl^-(aq)$

(c) $SO_3^{2-}(aq) + H_2O(l) \rightarrow SO_4^{2-}(aq) + 2H^+(aq) + 2e^-$

$Au^+(aq) + e^- \rightarrow Au(s)$

Electrolysis as a redox process

We dealt with **electrolysis** on pages 46–48. In one example we considered what happens when copper(II) chloride solution is electrolysed using a direct current (see Figure 14.11). During the course of this experiment, $Cu^{2+}(aq)$ ions are attracted to the negative electrode where they *gain* electrons and form copper atoms. This is a **reduction** reaction because it involves the *gain* of electrons.

$$Cu^{2+}(aq) + 2e^- \rightarrow Cu(s) \qquad \text{(reduction)}$$

At the same time, $Cl^-(aq)$ ions are attracted to the positive electrode where they *lose* electrons and form chlorine molecules. This is an **oxidation** reaction because it involves the *loss* of electrons.

$$2Cl^-(aq) \rightarrow Cl_2(g) + 2e^- \qquad \text{(oxidation)}$$

In fact, during any electrolysis process using a direct current, electron loss takes place at the positive electrode and electron gain at the negative electrode. Oxidation therefore takes place at the positive electrode where negative ions lose electrons, whereas reduction takes place at the negative electrode where positive ions gain electrons.

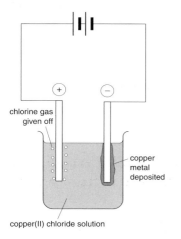

chlorine gas given off

copper metal deposited

copper(II) chloride solution

Figure 14.11 ▲
Electrolysis of copper(II) chloride solution.

Question

9 Write ion-electron equations for the reactions taking place at (i) the negative electrode and (ii) the positive electrode during the electrolysis of the following compounds using a direct current. In each case indicate whether the reaction is an example of oxidation or reduction.

(a) molten lead(II) bromide
(b) molten sodium iodide
(c) molten aluminium oxide

Reactions of metals

Some metals react very quickly and easily. For example, a freshly cut piece of sodium has a shiny, silvery appearance, but within seconds the surface loses its shiny appearance as gases in the air react with it. However, copper stays shiny for much longer before it too reacts with the gases around it and a green compound forms on its surface. You may have seen copper roofs on buildings that have turned green over a period of years. Gold is usually not affected by the gases in the air at all. The gold covering the dome of the mosque in Figure 14.12, for example, has not undergone any chemical reaction even after hundreds of years.

Figure 14.12 ▲
Why has the gold covered temple stayed shiny for so long?

From the observations above, we can put the three metals, sodium, copper and gold in order of reactivity. We can say that sodium is more reactive than copper which is more reactive than gold

A **reactivity series** of metals is a kind of 'league table of metals' which puts them in order of reactivity. Table 14.4 shows a common reactivity series for some common metals. In a reactivity series, the metals are placed in order of reactivity, with the most reactive at the top. The reactivity of the metals decreases as you go down the series. Notice that the order of metals in a reactivity series is very similar to that in the electrochemical series (ECS)

Metals can be placed in a reactivity series by observing their reactions with, for example, oxygen, water and dilute acids.

Reaction with oxygen

All of the metals above mercury in the reactivity series combine with oxygen when heated. This produces a metal oxide. The higher a metal is in the series, the more violent the reaction between the metal and oxygen.

Name	Symbol
potassium	K
sodium	Na
lithium	Li
calcium	Ca
magnesium	Mg
aluminium	Al
zinc	Zn
iron	Fe
tin	Sn
lead	Pb
copper	Cu
mercury	Hg
silver	Ag
gold	Au

Table 14.4 ▲
A reactivity series of metals.

The reactions of the metals copper to magnesium in the reactivity series can be studied using the apparatus shown on page 195. The general word equation is:

$$\text{metal} + \text{oxygen} \rightarrow \text{metal oxide}$$

For example:

$$\text{magnesium} + \text{oxygen} \rightarrow \text{magnesium oxide}$$
$$2Mg + O_2 \rightarrow 2MgO$$

Showing ions and state symbols:

$$2Mg(s) + O_2(g) \rightarrow 2Mg^{2+}O^{2-}(s)$$

Note: The three metals at the top of the reactivity series, potassium, sodium and lithium, are so reactive that they are stored under oil to prevent them from reacting with gases in the air such as oxygen, water vapour and carbon dioxide.

Reaction with water

The metals above aluminium in the reactivity series react with water to produce hydrogen gas and the corresponding metal hydroxide. Pieces of potassium, sodium and lithium move around on the surface of water as they react because they are less dense than water.

$$\text{metal} + \text{water} \rightarrow \text{metal hydroxide} + \text{hydrogen}$$

For example, sodium reacts with water to produce sodium hydroxide solution and hydrogen.

$$\text{sodium} + \text{water} \rightarrow \text{sodium hydroxide} + \text{hydrogen}$$
$$2Na + 2H_2O \rightarrow 2NaOH + H_2$$

Showing ions and state symbols:

$$2Na(s) + 2H_2O(l) \rightarrow 2Na^+(aq) + 2OH^-(aq) + H_2(g)$$

Reaction with dilute acids

All metals above copper in the reactivity series react with a dilute acid, such as hydrochloric acid or sulphuric acid, to produce a salt and hydrogen.

$$\text{metal} + \text{acid} \rightarrow \text{salt} + \text{hydrogen}$$

For example, zinc reacts with dilute hydrochloric acid to produce zinc(II) chloride solution and hydrogen.

$$\text{zinc} + \text{hydrochloric acid} \rightarrow \text{zinc(II) chloride} + \text{hydrogen}$$
$$Zn + 2HCl \rightarrow ZnCl_2 + H_2$$

Showing ions and state symbols:

$$Zn(s) + 2H^+(aq) + 2Cl^-(aq) \rightarrow Zn^{2+}(aq) + 2Cl^-(aq) + H_2(g)$$

Note: Potassium, sodium and lithium are too reactive to risk in reactions with acids.

Aluminium is unusual in its reaction with acids. It reacts only very slowly to begin with, but after about 20 minutes can be reacting very rapidly. The reason for this is that it is protected by a thin layer of aluminium oxide which must first be removed by the acid.

Question

10 Write (i) balanced formula equations and (ii) balanced equations showing state symbols and ions for the reactions between:

    (a) aluminium and oxygen
    (b) sodium and oxygen
    (c) lithium and water
    (d) calcium and water
    (e) magnesium and dilute sulphuric acid
    (f) aluminium and dilute hydrochloric acid

A summary of the reactions of metals

Metal	Oxygen	Reaction with Water	Dilute acid
potassium sodium lithium calcium magnesium	metal + oxygen ↓ metal oxide	metal + water ↓ metal hydroxide + hydrogen	metal + acid ↓ salt + hydrogen
aluminium zinc iron tin lead			
copper			
mercury silver gold	no reaction	no reaction	no reaction

Prescribed Practical Activity

Reaction of metals with oxygen

- The aim of this experiment is to place three metals, such as magnesium, zinc and copper, in order of reactivity by observing their reaction when heated in oxygen.

Continued ➤

- The apparatus used is shown.
- The metal is heated first, then the potassium permanganate and the metal are heated alternately, for a few seconds each, until a reaction involving the metal is observed.
- The higher a metal is in the reactivity series, the more vigorous is its reaction with oxygen.

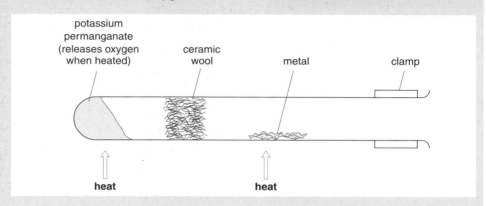

Extracting metals from their ores

How did the ancient craftsmen, who made the ornament shown in Figure 14.13, obtain the gold that it is made from? The answer is that they simply found it, either in the ground or in the bed of a stream. Gold is found as the metal element – it is not combined with other elements. Only unreactive metals like gold and silver are found uncombined in the Earth's crust.

The extraction of gold is quite easy in theory since it is usually mixed up with much lighter rocky or sandy material. This can be washed away with water to leave the much denser gold particles. You may have seen old photographs of prospectors 'panning' for gold. Unreactive metals, like gold and silver, were the first to be discovered because their extraction did not involve a chemical reaction.

Most metals are obtained from **ores** which are naturally occurring compounds from which the metal is extracted by chemical reactions. Many ores are oxides; haematite (iron(III) oxide) and bauxite (aluminium oxide) being two of the most important. Other ores that are not oxides can often be converted easily to the oxide. Galena (lead(II) sulphide) is an example of one such ore (see Table 14.5 and Figure 14.14).

Figure 14.13 ▲
A gold ornament made thousands of years ago.

Ore	Compound present
haematite	iron(III) oxide
bauxite	aluminium oxide
galena	lead(II) sulphide

Table 14.5 ▲
Some common ores.

Figure 14.14 ▲
Galena (lead(II) sulphide) crystals.

Extracting metals by heating

Reactive metals form stable compounds and so 'hold on' to oxygen more strongly than less reactive metals. Unreactive metals form *less* stable compounds and have a weaker hold on oxygen. For the oxides of metals below copper in the reactivity series, the metal's hold on oxygen is so weak that they can be decomposed to give the metal and oxygen by the use of heat alone.

$$\textbf{metal oxide} \rightarrow \textbf{metal} + \textbf{oxygen}$$

For example:

$$\text{mercury(II) oxide} \rightarrow \text{mercury} + \text{oxygen}$$

$$2HgO \rightarrow 2Hg + O_2$$

Extracting metals by heating with carbon or carbon monoxide

The oxides of metals above mercury in the reactivity series are not affected by heat alone. Several of their metal oxides can, however, be reduced to the metal by heating in the presence of another substance that is able to 'take away' the oxygen. Oxides of metals below aluminium in the reactivity series can be reduced to the metal by heating with carbon or carbon monoxide. These reactions take place because carbon bonds more strongly with oxygen than the metal does.

Heating with carbon:

$$\textbf{metal oxide} + \textbf{carbon} \rightarrow \textbf{metal} + \textbf{carbon dioxide}$$

For example:

$$\text{zinc(II) oxide} + \text{carbon} \rightarrow \text{zinc} + \text{carbon dioxide}$$

$$2ZnO + C \rightarrow 2Zn + CO_2$$

Heating with carbon monoxide:

$$\textbf{metal oxide} + \textbf{carbon monoxide} \rightarrow \textbf{metal} + \textbf{carbon dioxide}$$

For example:

$$\text{lead(II) oxide} + \text{carbon monoxide} \rightarrow \text{lead} + \text{carbon dioxide}$$

$$PbO + CO \rightarrow Pb + CO_2$$

Heating with carbon does *not* release the metal from oxides of reactive metals (those above zinc in the reactivity series). This is because the metal bonds more strongly with oxygen than carbon does.

A summary of the reactions of metal oxides

Metal	Effect of heating metal oxide	
	Alone	**With carbon or carbon monoxide**
potassium sodium lithium calcium magnesium aluminium	no reaction	no reaction
zinc iron tin lead copper mercury silver gold	metal oxide ↓ metal + oxygen	metal oxide + carbon or carbon monoxide ↓ metal + carbon dioxide

Extracting metals by electrolysis

All of the metals above zinc in the reactivity series are obtained by the electrolysis of molten compounds using a direct current. Aluminium is obtained from its oxide, but the rest are obtained from their chlorides. In all cases, once the ionic lattice is broken down and the ions become free to move, positive metal ions are attracted to the negative electrode where they gain electrons and form metal atoms. For example:

$$Al^{3+} + 3e^- \rightarrow Al$$

Industrial scale production of reactive metals has only been possible for just over a century because large amounts of electricity are needed for their extraction.

Extraction of iron in a blast furnace

Since the industrial revolution in the eighteenth century there has been a great demand for iron metal. It is needed for the construction of bridges, railways, trains, ships, lorries, cars, buildings and a host of other things.

Iron is extracted from iron ore, which is usually iron(III) oxide. The process is carried out in a huge structure called a blast furnace, which can be up to 70 metres high (see Figure 14.15). A mixture of iron ore, coke (which is mainly carbon) and limestone is added at the top of the furnace. At the same time blasts of hot air are blown in through small holes at the bottom. Oxygen in the blast of hot air reacts with the coke to form carbon dioxide.

$$\text{I} \qquad\qquad C + O_2 \rightarrow CO_2$$

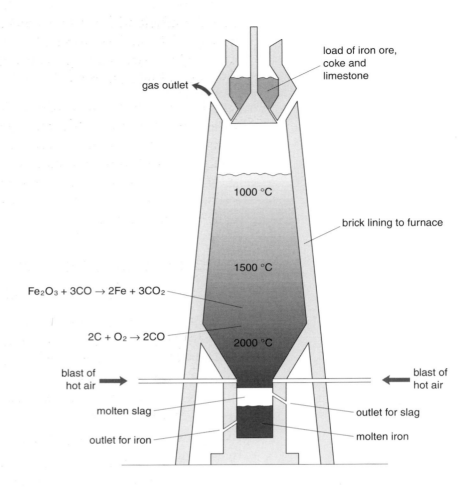

Figure 14.15 ▲
Inside a blast furnace.

As the carbon dioxide formed rises through the very hot coke they react to form carbon monoxide.

II $$C + CO_2 \rightarrow 2CO$$

Taking equations I and II together we can see that the carbon in the coke undergoes incomplete combustion to form carbon monoxide.

$$2C + O_2 \rightarrow 2CO$$

This exothermic reaction enables a temperature of about 2000°C to be maintained in the centre of the blast furnace providing the conditions needed for the successful extraction of iron. The carbon monoxide reduces the iron ore (iron(III) oxide) to iron.

$$Fe_2O_3 + 3CO \rightarrow 2Fe + 3CO_2$$

The high furnace temperature melts the iron. The molten iron runs to the bottom of the furnace where it is tapped from time to time.

Why is limestone used in the blast furnace?
The limestone is added to remove sandy impurities in the iron ore.

> Blast furnaces operate continuously, 24 hours a day. During that time one blast furnace can produce 10 000 tonnes of iron. Blast furnaces are designed to run for two years non-stop!

These combine with the limestone to form a molten 'slag' which floats on top of the molten iron and is run off separately. Once solidified, the 'slag' is processed to provide a range of useful products, including cement and insulation blocks for buildings.

Making steel from blast furnace iron

Blast furnace iron is brittle, mainly due to a high carbon content (about 5%). This can be reduced, and other non-metal impurities removed, by blowing oxygen through the molten iron. What is produced is a stronger, more flexible material known as mild steel. Mild steel has a carbon content of about 0.15%, the rest being almost entirely made up of iron.

Questions

11 Write balanced formula equations for each of the following:

    (a) Silver(I) oxide heated alone.

    (b) Tin(IV) oxide reduced to the metal by heating with carbon.

    (c) Copper(I) oxide reduced to the metal by heating with carbon monoxide.

    (d) Magnetic iron oxide, Fe_3O_4, reduced to the metal by heating with carbon monoxide.

12 (a) Why is it possible to extract zinc metal from zinc oxide by heating with carbon, but not possible to extract aluminium from aluminium oxide using the same procedure?

    (b) Why was the large scale extraction of reactive metals such as potassium, sodium and magnesium not possible until about a century ago?

13 (a) Name the three substances that are constantly fed into the top of a blast furnace and state why each is present.

    (b) Describe the sequence of three reactions inside a blast furnace that lead to the extraction of iron.

    (c) How are the high temperatures inside a blast furnace maintained?

Corrosion

Corrosion and rusting

Earlier in this section on page 194 we considered the different rates at which metals reacted with substances in the air. The silvery surface of a freshly cut piece of sodium goes dull within seconds. Copper stays shiny for much longer, but then a green substance gradually forms on its surface. Gold stays shiny for years and appears not to change. We could add to our observations that silvery grey iron becomes covered with a reddish-brown substance more quickly than copper turns green.

Figure 14.16 ▲
An example of well-advanced corrosion.

The metals sodium, iron and copper **corrode** when exposed to the atmosphere. In this simple example we can see that different metals corrode at different rates and that reactive metals corrode at a faster rate than less reactive metals. **Corrosion** is, in fact, a chemical reaction which involves the surface of a metal changing from an element to a compound. In extreme cases of corrosion, the metal can be eaten away completely. This can happen in the case of iron, for example, because the compound formed flakes away easily, constantly exposing fresh metal. The car in Figure 14.16 is an example of such rampant corrosion!

Since more iron is extracted and used than any other metal, it is perhaps reasonable that there should be a special word that describes the corrosion of iron. That word is **rusting** and the name given to the compound formed when iron corrodes is **rust**. Rusting is a huge problem world-wide that is very costly for many countries. This is because each year huge amounts of money must be used to replace rusted lorries, buses, cars, trains, bridges, ships, etc.

What conditions are needed for rusting to take place?

You may have noticed that iron objects appear to rust more quickly if they are wet. Iron nails left out in the rain go rusty more quickly than those left out in drier conditions. So we might wonder if both air and water are needed for rusting to take place.

In order to investigate what causes rusting, four identical iron nails are placed test tubes as shown in Figure 14.17. After a few days they are examined to see if they have rusted. Note that in tube 1, calcium chloride is present to remove any moisture from the air. In tube 2 the water has been boiled for several minutes to drive out any dissolved air. The layer of oil stops any air from dissolving back into the water.

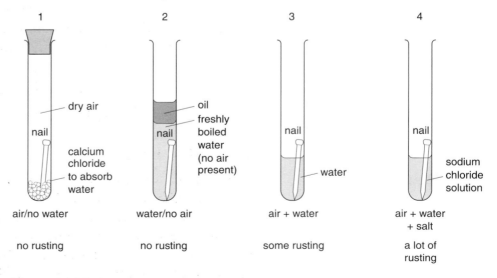

Figure 14.17 ▲
What causes iron nails to rust?

From this experiment it can be concluded that both air and water are needed for rusting to occur. Also, that common salt, sodium chloride, speeds up rusting.

Rusting, therefore, requires air, but which of the gases present in the air does it need? All of them, or just one? The experiment in Figure 14.17 shows that *oxygen* is needed for rusting to take place. As the moist iron filings rust, the water level in the test tube rises. It rises by about one-fifth of the height of the test tube – remember that oxygen makes up about one-fifth of the air.

From these experiments it may be concluded that both *water* and *oxygen* are needed for iron to rust.

Rusting as a redox process

The first stage in the rusting of iron is the loss of electrons by iron atoms to form iron(II) ions in solution, $Fe^{2+}(aq)$:

$$Fe(s) \rightarrow Fe^{2+}(aq) + 2e^- \qquad \text{(oxidation)}$$

In solution, the iron(II) ions lose another electron to form iron(III) ions, $Fe^{3+}(aq)$:

$$Fe^{2+}(aq) \rightarrow Fe^{3+}(aq) + e^- \qquad \text{(oxidation)}$$

The electrons lost are gained by water and oxygen forming hydroxide ions, $OH^-(aq)$:

$$4e^- + 2H_2O(l) + O_2(aq) \rightarrow 4OH^-(aq) \qquad \text{(reduction)}$$

When $Fe^{3+}(aq)$ ions and $OH^-(aq)$ ions meet, rust forms:

$$Fe^{3+}(aq) + 3OH^-(aq) \rightarrow Fe^{3+}(OH^-)_3(s)$$
$$\text{(rust)}$$

Note: Rust can also be looked upon as a form of iron(III) oxide called hydrated iron(III) oxide. When heated it loses water leaving anhydrous iron(III) oxide (anhydrous means 'without water'):

$$\left.\begin{array}{c} 2Fe(OH)_3 \\ \text{(rust)} \\ Fe_2O_3.3H_2O \end{array}\right\} \rightarrow Fe_2O_3 + 3H_2O$$

Detecting rusting

To find out if rusting is taking place, we can use **ferroxyl indicator**. This is a yellow solution that turns blue in the presence of iron(II) ions, $Fe^{2+}(aq)$, which are the ions formed when iron metal starts to rust. In general, the more blue colour there is, the more rusting has taken place. Ferroxyl indicator can also be used to detect fairly high concentrations of hydroxide ions, $OH^-(aq)$. If a solution is quite alkaline (about pH 10 or above), then ferroxyl indicator will turn pink.

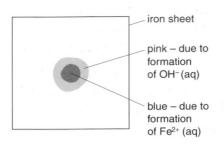

Figure 14.18 ▲
A drop of ferroxyl indicator on sheet iron.

When a drop of ferroxyl indicator is placed on a piece of sheet iron (in the form of mild steel sheet) blue and pink colours are soon formed (see Figure 14.18). The blue colour at the centre shows that that Fe^{2+}(aq) ions are being formed and therefore rusting is taking place. The pink outer colour shows that OH^-(aq) ions are being formed in that area.

At the centre of the drop iron atoms lose electrons to form Fe^{2+}(aq) ions. The electrons lost by the iron flow through the iron sheet to the edge of the drop where they join with oxygen and water molecules to form OH^-(aq) ions. A chemical cell has been set up on the iron.

Question

17 What colour does ferroxyl indicator give with:

   (a) Fe^{2+}(aq) ions
   (b) OH^-(aq) ions?

Speeding up corrosion

As we have seen already on page 200 common salt (sodium chloride) speeds up the rusting of iron. If you added common salt to the ferroxyl indicator used for the experiment in Figure 14.18, the colours would appear more quickly. All soluble salts would be expected to speed up the corrosion of metals. This is because they contain ions and therefore can act as electrolytes in the chemical cells that are set up when metals corrode. Common salt is spread on roads during winter and this speeds up the rusting of motor vehicles that use them.

Acid rain also speeds up the corrosion of metals. This is partly because the acids present can react with most metals to produce a salt and release hydrogen gas. It is also because acid solutions are ionic and therefore can act as electrolytes in the same way as soluble salts. The salts that are produced by the reaction between acids and metals can also act as electrolytes, speeding up corrosion.

Protection against corrosion

Physical protection

For a metal like iron to corrode it needs to be exposed to oxygen and water in the presence of an electrolyte. One way to prevent corrosion is therefore to put a barrier around the metal to keep out these substances. This is why the Forth Railway Bridge girders are painted. The paint gives physical protection to the steel, keeping out oxygen, water and electrolytes. Ships and vehicles are also given physical protection by means of paint.

Question

18 Why do salts speed up the corrosion of iron?

Physical barriers to corrosion
paint
oil or grease
plastic coating

Table 14.6 ▲

The moving parts of machines, such as a bicycle chain, cannot be protected by paint because it would be scratched off. Instead they are oiled or greased, which also helps with lubrication and reduces friction.

Metals can also be given physical protection by covering them with plastic. One way of achieving this is by dip-coating. The metal object is heated and then dipped into a powdered thermoplastic. Plastic powder in contact with the hot metal melts and sticks to it, then, as the metal cools, the plastic solidifies.

Other metals, such as zinc, tin and chromium can be used to provide physical protection by forming a layer on another metal. All three can be deposited as a thin layer by means of electrolysis in a process called electroplating (see page 207). Zinc coatings can also be achieved by dipping the metal into molten zinc.

(see page 207)

Figure 14.19 ▲
The Forth Railway Bridge.

Direct electrical protection

When a metal corrodes to form a compound, metal atoms lose electrons and form ions. Therefore, if electrons could be forced back onto the metal it would not corrode. Electrons can be supplied from the negative terminal of any direct current supply such as a battery.

If two iron nails are dipped into ferroxyl indicator solution and connected by a battery (Figure 14.20) a blue colour forms around the positive electrode and a pink colour forms around the negative electrode. Electrons are taken away from the nail attached to the positive terminal of the battery and it corrodes forming Fe^{2+}(aq) ions (which give a blue colour with ferroxyl indicator). Electrons are given to the nail attached to the negative terminal. This nail is protected from corrosion, but the electrons are gained by oxygen and water, producing OH^-(aq) ions (which give a pink colour with ferroxyl indicator).

Question

19 What is meant by the phrase 'physical protection from corrosion'?

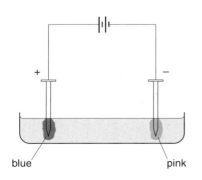

Figure 14.20 ▲
Two iron nails, joined by a battery, dipping into ferroxyl indicator solution.

This method of protection is widely used in motor vehicles where the negative terminal of the battery is connected to the steel bodywork (see Figure 14.21). Unfortunately, although corrosion of vehicles is slowed down, it is not stopped completely.

Many piers and jetties and even some ocean liners, when they are in port, use direct electrical protection to reduce corrosion.

Figure 14.21 ▲
Attaching the negative terminal of the battery to the car body reduces the rate of corrosion.

Question

20 What would be the effect of attaching the *positive* terminal of a car battery to the steel body of a car. Explain your answer.

Sacrificial protection

You have already learned that all metals tend to lose electrons. You also know that if two metals are joined in a simple chemical cell, electrons will flow from the metal that is higher in the ECS to the one lower down. What this means in terms of corrosion is that a metal higher in the ECS will corrode to protect the one lower down if they are in contact, or if there is a wire joining them to allow electrons to flow. This is called **sacrificial protection** because the higher metal is sacrificed to protect the lower metal.

In the simple cell shown in Figure 14.22, electrons flow from magnesium to iron through the wires and meter. This happens because magnesium is higher in the ECS than iron and loses electrons more readily. As a result, the magnesium corrodes and is sacrificed, but the iron is protected. The reactions taking place are:

At the magnesium:

$$Mg(s) \rightarrow Mg^{2+}(aq) + 2e^-$$

At the iron (and at the magnesium):

$$4e^- + 2H_2O(l) + O_2(aq) \rightarrow 4OH^-(aq)$$

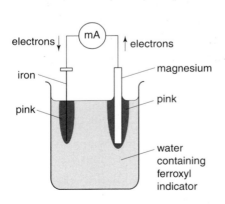

Figure 14.22 ▲
In this simple cell electrons flow from magnesium (sacrificed) to iron (protected).

Attaching a metal that is higher in the ECS protects iron from corrosion, but attaching one that is lower down speeds up its corrosion. We can see this if we take three iron nails, with magnesium attached to one and copper to another, that are placed in ferroxyl indictor solution. Figure 14.23 gives a view from above three shallow dishes that contain these experiments.

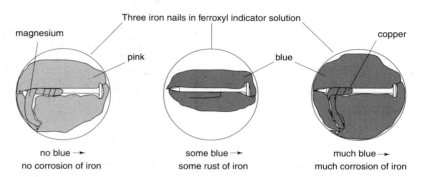

Attaching copper to iron speeds up the corrosion of iron. This happens because iron is higher in the ECS than copper. Under corroding conditions, electrons therefore flow from the iron to the copper. Iron is sacrificed to protect the copper. The result can be verified by setting up a copper/iron cell as shown in Figure 14.24. The colours produced are explained by examining the reactions taking place at each metal:

At the iron: $Fe(s) \rightarrow Fe^{2+}(aq) + 2e^-$

At the copper: $4e^- + 2H_2O(l) + O_2(aq) \rightarrow 4OH^-(aq)$

Sacrificial protection is widely used in the constant battle against corrosion. For example, bags of scrap magnesium are attached at intervals to steel pipelines (see Figure 14.25) and zinc blocks are attached to the steel hulls of ships. Remember that steel is mainly iron. In both cases, under corroding conditions, electrons flow from the magnesium or zinc to the iron in the steel. This is because both magnesium and zinc are above iron in the ECS. The result is that the magnesium and zinc are sacrificed to protect the iron and must, from time to time, be replaced.

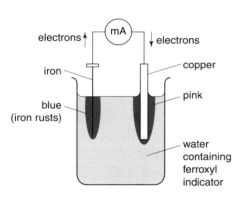

Figure 14.24 ▲
In this simple cell electrons flow from the iron (sacrificed) to the copper (protected).

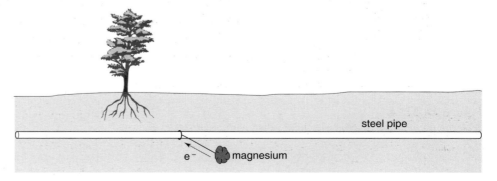

Figure 14.25 ▶
Magnesium supplies electrons to the iron in the steel pipe, protecting it by sacrificial corrosion. Ions in the damp soil complete the circuit.

Questions

21 Explain fully why zinc blocks are attached to the steel (mainly iron) hulls of ships.

22 1p and 2p coins are made of copper-plated steel. Explain the corrosion implications of a break in the copper coating of one such coin

Galvanised iron

Iron and steel objects can be given a protective coating of zinc by dipping them in the molten metal. The process is called **galvanising** and the product is referred to as **galvanised** iron or steel. Galvanising provides iron and steel with 'double' protection. Firstly, the zinc coating is tough and provides *physical* protection by keeping out oxygen and water. Secondly it can provide *sacrificial* protection if the coating is broken. The zinc can protect the iron, even when the zinc coating is broken because zinc is above iron in the ECS and so, under corroding conditions, electrons flow from the zinc to the iron. The zinc is sacrificed to protect the iron (see Figure 14.26).

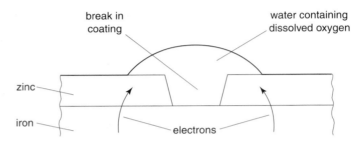

Figure 14.26 ▲
Galvanised iron is protected by zinc even with the coating broken.

A wide variety of iron and steel objects can be galvanised, including nails, motorway crash barriers and car bodies. The steel of the Forth Road Bridge was galvanised in a special way – 500 tonnes of molten zinc was sprayed on to it.

Figure 14.27 ▲
Both the car and the bridge were made using galvanised steel.

Tin-plated iron

Tin-plated steel (mainly iron) is widely used for items such as food cans and biscuit tins. Steel strip is given a thin coating of tin by electroplating (dealt with next). Tin has a shinier appearance than zinc and is less reactive. The lower activity means that it provides good *physical* protection for the steel, keeping oxygen and water out. However, if the tin coating is broken under corroding conditions the iron in the steel rusts very rapidly. The reason for this is that iron is above tin in the ECS and therefore electrons flow from the iron to the tin. The result of this is that the iron is sacrificed to protect the tin (see Figure 14.28).

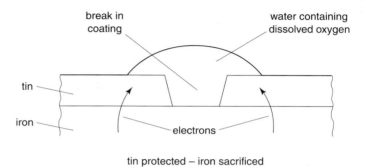

Figure 14.28 ▲
When the tin coating on tin-plated iron is broken, iron corrodes to protect the tin.

Questions

23 Explain what colours you would expect to form if ferroxyl indicator was added to the water shown in Figures 14.26 and 14.28.

24 When iron rusts it first loses electrons forming iron(II) ions. The electrons that are lost by iron are gained by oxygen and water. Write ion-electron equations for these reactions.

25 Which ions does ferroxyl indicator detect and what colours are produced by each ion?

26 Explain how galvanising provides iron with both physical and chemical protection.

27 Explain why, when the tin layer on tin-coated iron is broken, the rusting of the iron can take place rapidly.

Electroplating

If you look on the back of a spoon you might find the letters 'epns' if the spoon is silver-plated. The letters stand for '**e**lectro**p**lated **n**ickel **s**ilver' because a nickel spoon has been plated with silver using electrolysis. During electroplating the object to be plated is used as the negative electrode in a solution containing ions of the metal to be deposited.

The positive electrode is usually made of the same metal as that being deposited at the negative electrode – this maintains the concentration of the metal ions in solution. The diagram in Figure 14.29 shows how this can be done in the laboratory.

The tin-plating of steel is carried out by passing steel strip through a solution containing tin(II) ions, $Sn^{2+}(aq)$. While it is in the solution, the steel becomes the negative electrode and a thin layer of tin is deposited on it. The reaction taking place at the steel electrode is as follows:

$$Sn^{2+}(aq) + 2e^- \rightarrow Sn(s)$$

Other metals that can be obtained as electroplatings include gold, chromium and zinc. Chromium-plating produces a very hard surface.

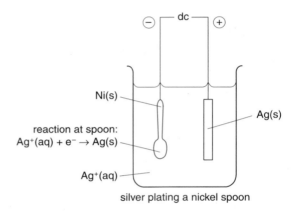

silver plating a nickel spoon

Figure 14.29 ▲
Silver plating a nickel spoon.

Chromium-plated parts are used in the engines of helicopters, for example.

Electroplating always provides physical protection from corrosion for the under-lying metal. Whether the plated metal provides sacrificial protection as well, when the coating is broken, depends upon the relative positions of the metals in the ECS. As we have already seen, tin-plating on iron provides physical protection, but not sacrificial protection when the coating is broken. This is because tin is below iron in the ECS.

In questions 1–4 choose the correct word **from the following list** to complete the sentence.

lower, gained, less, acids, more, higher, water, lost

1 In a simple chemical cell, electrons flow from the metal that is _____ in the ECS to the other metal.

2 The position of hydrogen in the ECS can be determined by the reaction of metals with _____.

3 During a reduction reaction electrons are _____ by a substance.

4 A metal that is higher in the reactivity series forms compounds that are _____ easily decomposed than a metal lower down.

5 Which of the following is an example of a redox reaction?

A $Ba^{2+}(aq) + SO_4^{2-}(aq) \rightarrow Ba^{2+}SO_4^{2-}(s)$
B $2H_2O(l) \rightarrow O_2(g) + 4H^+(aq) + 4e^-$
C $Fe(s) + 2H^+(aq) \rightarrow Fe^{2+}(aq) + H_2(g)$
D $2H^+(aq) + CO_3^{2-}(aq) \rightarrow H_2O(l) + CO_2(g)$

6 Dipping two metals into a salt solution and connecting them by means of wires and a voltmeter creates a simple cell. Which of the following pairs produce the **lowest** voltage?

A Tin and lead
B Aluminium and silver
C Zinc and copper
D Iron and lead

7 X, Y and Z are three metals. X reacts with dilute acid but not with water. Y does not react with dilute acid or water. Z reacts with water and dilute acid. The order of reactivity, starting with the most reactive is

A YXZ   B ZYX   C XZY   D ZXY

8 When iron is galvanised it is coated with:

A copper
B zinc
C tin
D lead

9 Which of the following shows the correct colours produced by ferroxyl indicator with the ions $Fe^{2+}(aq)$ and $OH^-(aq)$?

	Fe^{2+}(aq)	OH^-(aq)
A	pink	yellow
B	pink	blue
C	yellow	blue
D	blue	pink

10 In which of the following cases will displacement take place?

A Zinc and magnesium chloride solution
B Iron and tin(II) chloride solution
C copper and sodium sulphate solution
D silver and copper(II) sulphate solution

11 When molten potassium iodide is electrolysed, potassium forms at negative electrode and iodine at the positive electrode. Write ion-electron equations for the reactions taking place at the two electrodes, adding the words 'oxidation' and 'reduction' as appropriate.

12 The volume of gas released from a reaction between magnesium and dilute hydrochloric acid was noted at time intervals. The results are shown in the table below.

Time/s	Volume of gas/cm³
0	0
30	20.0
60	30.0
90	36.0
120	40.5
150	44.0
180	46.0
210	47.5
240	49.0
270	49.7
300	50.0

(a) Draw a line graph using these results.
(b) Give a balanced formula equation for the reaction.

13 Explain why attaching zinc blocks to the iron hulls of ships helps to stop the corrosion of iron.

Whole Course Summaries and Questions

15 Types of Chemical Formulae

Molecular formula shows the number of atoms of the different elements which are present in one molecule of a substance, e.g. C_2H_4 for ethene.

Empirical formula shows the simplest whole number ratio of atoms in a compound, e.g. SiO_2 for silicon dioxide.

Ionic formula shows the simplest whole number ratio of ions in a compound, e.g. $Ca^{2+}(Cl^-)_2$ for calcium chloride.

Full structural formula shows all the of the bonds in a molecule or ion but does not necessarily show the true shape.

e.g. a) methane

$$\begin{array}{c} H \\ | \\ H-C-H \\ | \\ H \end{array}$$

e.g. b) the ammonium ion

$$\left[\begin{array}{c} H \\ | \\ H-N-H \\ | \\ H \end{array}\right]^+$$

Shortened structural formula shows the sequence of groups of atoms in a molecule, e.g. $CH_3CH_2CH_2CH_3$ for butane.

Perspective formula shows the true shape of a molecule or ion.

e.g. a) methane is tetrahedral

e.g. b) water is angular (bent)

Note: What do you write when asked for 'the formula' or 'the simple formula' of a substance?

(a) For a **covalent molecular substance** we usually give the molecular formula, e.g. C_2H_6 for ethane. However sometimes the empirical formula is used instead, e.g. P_2O_3 for phosphorus oxide when in fact the molecular formula is P_4O_6.

(b) For an **ionic** or **covalent network compound** we usually write the empirical formula, but show no charges in the ionic compound, e.g. NaCl for sodium chloride and SiO_2 for silicon dioxide.

(c) For **elements** a single symbol is usually written for all apart from those which exist as diatomic molecules (hydrogen, oxygen, nitrogen and the halogens), e.g. C for carbon and Fe for iron, but H_2 for hydrogen and O_2 for oxygen, etc. (Only the noble gases exist as individual atoms not bonded to other atoms. In all other cases where single symbols are written as chemical formulae for elements, the symbol represents a sort of empirical formula.)

Chapter 15 Study Questions

In questions 1–4 choose the correct word(s) **from the following list** to complete the sentence.

covalent, full, molecular, shortened, ionic, network

1 C_5H_{12} is the _____ formula for pentane.

2 Covalent _____ compounds, such as silicon dioxide are usually represented by an empirical formula.

3 The _____ structural formula for propan-1-ol is $CH_3CH_2CH_2OH$.

4 The _____ structural formula for ammonia is

$$H\text{—}N\text{—}H$$
$$|$$
$$H$$

5 The full structural formula for oxalic acid is shown below.

(a) What is the molecular formula for oxalic acid?
(b) What is the empirical formula for oxalic acid?

6 Boron nitride is a covalent network compound that obeys the normal valency rules. Give the formula for this compound.

7 Phosphine, which has molecular formula PH_3, has the same molecular shape as ammonia, NH_3.

(a) Draw a full structural formula for phosphine.
(b) Draw a perspective formula for phosphine and state what word describes this shape.

8 Phenylethene, the monomer that is used to make the polymer poly(phenylethene), has the following formula.

The phenyl group, shown below, represents the group C_6H_5.

(a) Give the molecular formula for phenylethene.
(b) Give the empirical formula for phenylethene.

9 Trichloromethane (chloroform), which has molecular formula $CHCl_3$, was used many years ago as an anaesthetic.

(a) Draw a full structural formula for trichloromethane.
(b) Draw a perspective formula for trichloromethane.

10 For each of the following give i) the molecular formula, ii) the empirical formula, iii) the full structural formula and iv) the shortened structural formula.

(a) hexane
(b) but-2-ene
(c) ethanol
(d) propanoic acid

11 The simple formula for calcium chloride is $CaCl_2$. The ionic formula for the same compound is $Ca^{2+}(Cl^-)_2$. Rewrite the following simple formulae as ionic formulae.

(a) KNO_3
(b) AlF_3
(c) Na_2SO_4
(d) NH_4Cl
(e) $(NH_4)_3PO_4$

12 Draw perspective formulae to show the true shapes of the following molecules.

(a) hydrogen chloride
(b) hydrogen sulphide
(c) ethane
(d) tetrachlorosilane, $SiCl_4$

13 What formulae would you write for the following elements if they were to appear uncombined in a formula equation?

(a) sulphur
(b) bromine
(c) hydrogen
(d) magnesium
(e) xenon
(f) oxygen
(g) boron

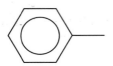

16 Types of Chemical Reaction

Addition The joining of two or more molecules to produce a single larger molecule and nothing else.

Addition polymerisation A reaction during which many small monomer molecules join to form one large polymer molecule and nothing else.

Burning A reaction in which a substance combines with oxygen releasing energy in the form of heat and light.

Catalytic hydration An addition reaction in which water adds on to a molecule in the presence of a catalyst.

Combustion See **burning**.

Condensation polymerisation A reaction during which many small monomer molecules join to form one large polymer molecule, with water or some other small molecule formed at the same time.

Corrosion A reaction in which the surface of a metal changes from an element to a compound.

Cracking The breaking up of larger hydrocarbon molecules (usually alkanes) to produce a mixture of smaller molecules (usually alkanes and alkenes). The use of a catalyst (in **catalytic cracking**) allows the reaction to take place at a lower temperature, thus saving energy and making the reaction more economical.

Decomposition The breaking down of a compound into two or more simpler substances (usually by heating).

Dehydration The removal of water from a compound. (Concentrated sulphuric acid is a powerful dehydrating agent.)

Displacement The formation of a metal from a solution containing its ions by reaction with a metal higher in the electrochemical series (ECS).

Fermentation The breakdown of glucose to produce ethanol and carbon dioxide catalysed by enzymes in yeast. (Can be applied to the similar breakdown of other organic molecules.)

Hydration An addition reaction in which water adds on to a molecule.

Hydrolysis A reaction in which a larger molecule is broken down into two or more smaller molecules by reaction with water. (All digestion reactions are examples of hydrolysis.)

Neutralisation A reaction of acids with bases in which the pH of a solution moves towards 7.

Oxidation A reaction in which electrons are lost. (They appear on the *right* hand side of the equation.)

Polymerisation A reaction during which many small monomer molecules join to form one large polymer molecule. (See **Addition polymerisation** and **condensation polymerisation**.)

Precipitation A reaction in which two solutions react to form an insoluble solid called a precipitate.

Redox A reaction in which both reduction and oxidation take place. Electrons are lost by one substance and gained by another.

Reduction A reaction in which electrons are gained. (They appear on the *left* hand side of the equation.)

Rusting A reaction during which *iron* corrodes and the surface of the metal changes into a compound.

Oxidation
Is
Loss of electrons

Reduction
Is
Gain of electrons

Note: Some reactions can be classified as being of more than one type. For example the reaction between solutions of an acid and an alkali may result in the formation of a precipitate. The reaction is therefore classified as both a neutralisation reaction and a precipitation reaction.

Chapter 16 Study Questions

In questions 1–4 choose the correct word(s) **from the following list** to complete the senetence.

> **addition, condensation, lost, lower, hydrolysis, gained, displacement, higher**

1 Catalytic hydration is an example of a type of reaction that is referred to as _____.

2 The reaction between ethanol and methanoic acid to form ethyl methanoate and water is an example of a _____ reaction.

3 During a reduction reaction electrons are _____.

4 During a displacement reaction one metal displaces another by reacting with ions of a metal that is _____ in the electrochemical series.

Questions 5–7 refer to the following types of reaction.

    A Cracking
    B Combustion
    C Neutralisation
    D Precipitation

To which of the above types of reaction do the following belong?

5 $CuO(s) + H_2SO_4(aq) \rightarrow CuSO_4(aq) + H_2O(l)$

6 $C_{13}H_{28}(l) \rightarrow C_9H_{20}(l) + C_4H_8(g)$

7 $CuSO_4(aq) + BaCl_2(aq) \rightarrow BaSO_4(s) + CuCl_2(aq)$

8 Name the type of chemical reaction in each of the following cases. In some cases more than one descriptive term may apply.

(a) $C_3H_6(g) + Br_2(aq) \rightarrow C_3H_6Br_2(l)$

(b) $Zn(s) + 2Ag^+(aq) \rightarrow Zn^{2+}(aq) + 2Ag(s)$

(c) $C_6H_{12}O_6(aq) \rightarrow 2C_2H_5OH(aq) + 2CO_2(g)$

(d) $MnO_4^-(aq) + 8H^+(aq) + 5e^- \rightarrow$
$\qquad\qquad\qquad\qquad Mn^{2+}(aq) + 4H_2O(l)$

(e) $CH_3COOC_2H_5 + H_2O \rightleftharpoons$
$\qquad\qquad\qquad\qquad CH_3COOH + C_2H_5OH$

(f) $nH_2N(CH_2)_5COOH \rightarrow$
$\qquad\qquad\qquad\qquad (HN(CH_2)_5CO\,)_n + nH_2O$

(g) $2C_6H_{12}O_6 \rightarrow C_{12}H_{22}O_{11} + H_2O$

(h) $2I^-(aq) \rightarrow I_2(s) + 2e^-$

(i) $nCH_2{=}CHCN \rightarrow (CH_2CHCN)_n$

(j) $3LiOH(aq) + H_3PO_4(aq) \rightarrow$
$\qquad\qquad\qquad\qquad Li_3PO_4(s) + 3H_2O(l)$

(k) $C_6H_{12}O_6(aq) \rightarrow 2C_2H_5OH(aq) + 2CO_2(g)$

(l) $nC_6H_{12}O_6 \rightarrow (C_6H_{10}O_5)_n + nH_2O$

Gas Production and Collection

Gas production

Several gases can be produced using a solid and a solution in the apparatus shown in Figure 17.1. Solid **A** is placed in the flask and solution **B** is added via the thistle funnel.

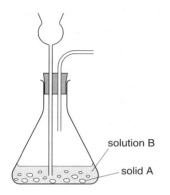

Figure 17.1 ▲

solution B

solid A

Solid A	Solution B	Gas produced
zinc	dilute sulphuric acid	hydrogen
marble chips ($CaCO_3$)	dilute hydrochloric acid	carbon dioxide
manganese(IV) oxide	hydrogen peroxide	oxygen

Gas collection

Collection over water

If a gas is *not* very soluble in water then it may be collected over water as shown in Figure 17.2. Gases that can be collected in this way include oxygen, nitrogen, carbon dioxide, the noble gases and all of the hydrocarbon gases such as methane.

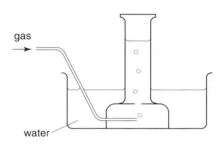

gas

water

Figure 17.2 ▲

Gases that are appreciably soluble in water and therefore *cannot* be collected over water include ammonia, sulphur dioxide and nitrogen dioxide. It should also be noted that collection over water does not provide a *dry* sample of gas.

Collection by downward displacement of air

Gases that are *less* dense than air may be collected by downward displacement of air as shown in Figure 17.3. Since air is roughly 80% nitrogen, which has a formula mass of 28, and 20% oxygen, with a formula mass of 32, the average air molecule has a formula mass of about 29. If a gas has a formula mass that is less than 29 it is less dense than air and can be collected by downward displacement of air.

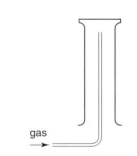

gas

Figure 17.3 ▲

Gases that are appreciably less dense than air, and therefore can be collected using this method include hydrogen, helium, ammonia and methane. Collection by this method provides a *dry* sample of gas, but since none of the lighter gases are coloured, it is difficult to know when the gas jar or test tube is full.

Collection by upward displacement of air

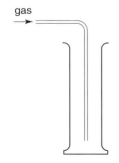

gas

Figure 17.4 ▲

Gases that are *more* dense than air may be collected by upward displacement of air as shown in Figure 17.4. Examples of gases that have formula masses well above 29 and can therefore be collected using this method include carbon dioxide, sulphur dioxide, nitrogen dioxide and butane.

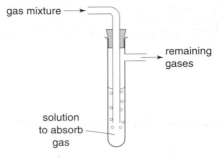

gas mixture →

→ remaining gases

solution to absorb gas

Figure 17.5 ▲

Collection by this method provides a *dry* sample of gas, but only in the case of a coloured gas, such as the reddish brown nitrogen dioxide, is it easy to tell when the gas jar or test tube is full.

Removing acidic and alkaline gases

An acidic gas, such as carbon dioxide, can be removed from other neutral gases by passing the mixture through a solution of an alkali such as sodium hydroxide. An alkaline gas, such as ammonia, can be removed from other neutral gases by passage through an acid solution, such as dilute hydrochloric acid. Apparatus that could be used to remove acidic or alkaline gases is shown in Figure 17.5.

Although ammonia is the only common alkaline gas, there are several fairly common acidic gases. Examples of acidic gases include sulphur dioxide, nitrogen dioxide and hydrogen chloride.

Chapter 17 Study Questions

In questions 1–4 choose the correct word(s) **from the following list** to complete the sentence.

downward, less, more, upward, nitrogen dioxide, nitrogen, insoluble, soluble

1 Neon is _____ dense than air and can be collected by downward displacement of air.

2 Ammonia is _____ in water and is therefore collected by downward displacement of air.

3 Carbon dioxide can be collected by _____ displacement of air.

4 Because _____ is insoluble in water it can be collected over water.

5 Which of the following gases is more dense than air?

A Ammonia
B Methane
C Helium
D Argon

6 Which of the following gases can be collected over water?

A Ammonia
B Carbon dioxide
C Nitrogen dioxide
D Sulphur dioxide

7 Which of the following can be collected by downward displacement of air?

A Krypton
B Xenon
C Argon
D Helium

8 Oxygen, nitrogen, hydrogen and carbon dioxide are common gases.

A All can be collected by upward displacement of air.
B All can be collected by downward displacement of air.
C All can be collected over water.
D Different methods must be used for their collection.

9 A sample of argon gas (neutral and insoluble) is contaminated with acidic hydrogen bromide gas. Copy and complete the diagram below to show how the hydrogen bromide could be removed and the argon collected over water.

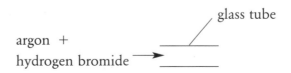

glass tube

argon + hydrogen bromide →

Filtration

- Insoluble solids can be separated from liquids by passage through filter paper.
- The solid which remains behind in the filter paper is called the **residue**.
- The liquid which passes through the filter paper is called the **filtrate**.
- Precipitates may be separated from aqueous solutions of other substances using this technique.

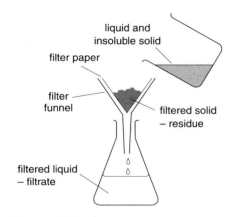

Figure 18.1 ▲
Apparatus for filtration.

Evaporation

- A solid solute may be recovered from a solution by evaporating off the solvent.
- Soluble salts may be recovered from solutions using this technique.

Distillation

- A mixture of liquids may be separated by distillation, the one with the lowest boiling point distilling off first.
- Ethanol (boiling point 79°C) may be separated from water (boiling point 100°C) after the fermentation of sugars such as glucose.
- A solvent can be recovered from a solution containing dissolved solids.

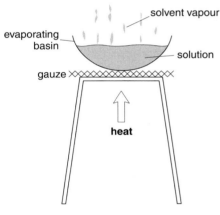

Figure 18.2 ▲
Apparatus for evaporation.

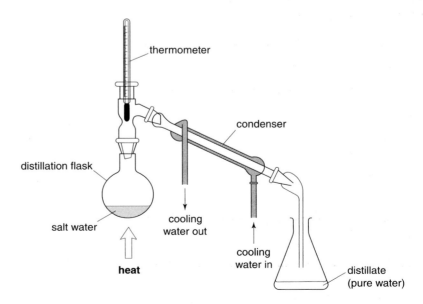

Figure 18.3 ▲
Apparatus for distillation.

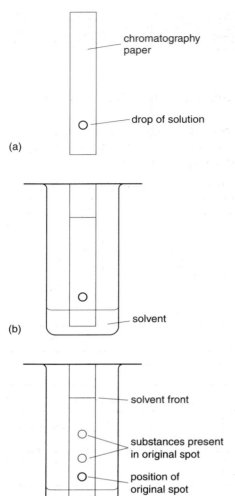

(a)

(b)

(c)

Figure 18.4 ▲
Paper chromatography can work using only a single drop of a solution.

Chromatography

- Small quantities of soluble compounds can be separated using paper chromatography.
- The separation of coloured substances, such as food colours or coloured inks, is easily seen.
- A drop of the mixture is spotted onto chromatography paper (or filter paper) and the bottom of the paper is placed in a suitable solvent (water can often be used).
- If the substances used are colourless, such as sugars or amino acids, a locating agent is used to show up their presence.

Whole Course Questions

Part I Grid questions

1 The diagram shows part of the Periodic Table. The letters do **not** represent the symbols for the elements.

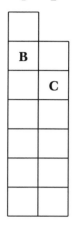

Group   1   2   3   4   5   6   7   0

(a) Identify the alkali metal.
(b) Identify the two elements that are in the same period.
(c) Identify the two elements that are gases at room temperature.

2 Shiraz was studying reaction speeds. In one of his investigations, he measured the time taken for 0.1 g of magnesium to completely react with hydrochloric acid under different conditions.

REACTION	A	B	C	D	E	F
Concentration of acid in mol/l	1.5	1.0	0.5	0.5	1.0	1.5
Temperature/°C	20	20	10	10	10	20
Form of magnesium	ribbon	ribbon	powder	ribbon	ribbon	powder

(a) Identify the reaction which was finished in the shortest time.
(b) Identify the **two** reactions which would allow Shiraz to examine the effect of temperature on reaction speed.

3 Many chemical compounds contain ions.

A	BaO	B	CaBr$_2$	C	CuO
D	KCl	E	NaBr	F	SrCl$_2$

(a) Identify the compound which gives a lilac flame colour.
(b) Identify the **two** compounds which are bases.
(c) Identify the compound in which **both** ions have the same electron arrangement as argon.

4 The table gives information about some atoms.

Atom	Atomic number	Mass number
P	22	50
Q	24	50
R	24	54
S	26	54
T	26	56

Identify the correct statement(s).

A	Atoms P and Q have the same number of protons.
B	Atoms Q and R have the same number of electrons.
C	Atoms P and S have the same number of neutrons.
D	Atoms R and S are isotopes of each other.
E	Atoms S and T have different chemical properties.

5 Each box in the grid below shows the formula of a substance.

A	CH$_4$	B	H$_2$S	C	N$_2$
D	O$_2$	E	CaCl$_2$	F	NH$_3$

(a) Identify the ionic compound.
(b) Identify the substance that exists as tetrahedral molecules.
(c) Identify the substance that contains double covalent bonds.

6 Six common elements from the Periodic Table are shown below.

A	lithium	B	sulphur	C	magnesium
D	potassium	E	nitrogen	F	oxygen

(a) Identify the element(s) that form(s) ions with the same electron arrangement as argon.

(b) Identify the element that forms ions with a double positive charge.

(c) Identify the **two** diatomic elements.

7 Neon is used in lighting and advertising displays. Identify the statement(s) which can refer to an atom of neon.

A	The atom has the same electron arrangement as a Cl⁻ ion.
B	The atom has two more electrons than an atom of oxygen.
C	The atom has the same number of outer electrons as an atom of helium.
D	The atom has an unstable electron arrangement.
E	The atom has the same number of electrons as an Al^{3+} ion.

8 The properties of substances are related to bonding.

Substance	Melting Point/°C	Boiling Point/°C	Ability to conduct as	
			a solid	a liquid
A	2617	4607	yes	yes
B	782	1600	no	yes
C	81	218	no	no
D	−183	−164	no	no
E	6	80	no	no
F	2300	2550	no	no

(a) Identify the substance which is a solid at room temperature (20°C) and would melt if placed in a test tube surrounded by boiling water.

(b) Identify the substance with ionic bonding.

(c) Identify the substance with a covalent network.

9 Electricity is passed through copper(II) dichromate solution.

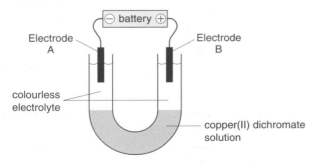

Copper(II) dichromate solution contains blue copper ions and orange dichromate ions.

Identify the correct statement(s).

A	Copper forms at electrode A.
B	The electrolyte around electrode A remains colourless.
C	A blue colour moves to electrode B.
D	Dichromate ions move to electrode B.
E	Electrons move through the solution from A to B.

10 Distillation of crude oil produces several fractions.

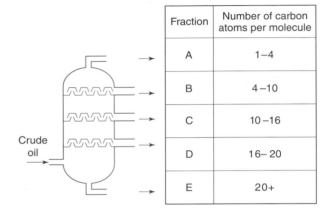

Fraction	Number of carbon atoms per molecule
A	1–4
B	4–10
C	10–16
D	16–20
E	20+

(a) Identify the fraction which is used as a fuel for jet aircraft.

(b) Identify the fraction with the lowest boiling point.

11 Mr Sharp demonstrated the fractional distillation of crude oil to class 3B.

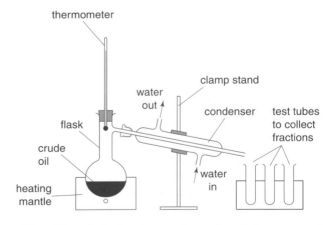

He collected six fractions and numbered them according to the order in which they were collected.

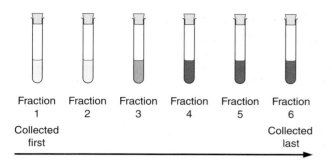

Fraction 1 — Collected first
Fraction 6 — Collected last

Identify the true statement(s).

A	Fraction 6 evaporates most easily.
B	Fraction 4 is more viscous than fraction 3.
C	Fraction 2 is more flammable than fraction 1.
D	Fraction 5 has a lower boiling range than fraction 4.
E	The molecules in fraction 3 are larger than those in fraction 2.

12 Hydrocarbons contain only atoms of carbon and hydrogen.

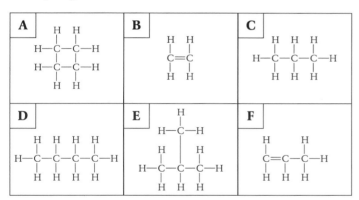

(a) Identify the hydrocarbon which reacts with hydrogen to form propane.
(b) Identify the hydrocarbon with the general formula C_nH_{2n} which does **not** immediately decolourise bromine solution.
(c) Identify the **two** isomers.

13 Alkanones and alkanoic acids are two families of carbon compounds. Each family has a particular arrangement of atoms.

alkanone   alkanoic acid

☐ represents the rest of the molecule

Alkanones can be prepared from alkanoic acids.

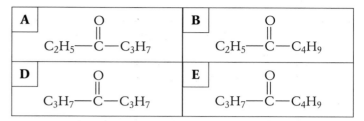

The grid shows the structural formulae for some alkanones.

A	C_2H_5—C(=O)—C_3H_7	B	C_2H_5—C(=O)—C_4H_9
D	C_3H_7—C(=O)—C_3H_7	E	C_3H_7—C(=O)—C_4H_9

(a) Identify the alkanone which can be produced by heating a mixture of:

$$C_2H_5\text{—C(=O)—OH} \quad \text{and} \quad C_4H_9\text{—C(=O)—OH}$$

(b) Identify the alkanone which can be prepared by heating only one alkanoic acid.

14 Proteins and fats are hydrolysed during digestion.

A	$C_{17}H_{35}$—C(=O)(OH)	B	CH_3—C(=O)—OC_2H_5	C	CH_2—CH—CH_2 with OH, OH, OH
D	C_2H_5—C(=O)(OH)	E	C_3H_7—NH_2	F	H_2N—CH_2—C(=O)(OH)

(a) Identify the compound which could be produced by the hydrolysis of a protein.
(b) Identify the compound(s) which could be produced by the hydrolysis of a fat.

15 Many different compounds are associated with foods.

A	glucose, a carbohydrate	B	glycerol, an alcohol	C	starch, a carbohydrate
D	glycine, an amino acid	E	oleic acid, a fatty acid	F	glyceryl tristearate, a fat

(a) Identify the polymer.
(b) Identify the compound which could be formed by the hydrolysis of a protein.
(c) Identify the **two** compounds which could be formed by the hydrolysis of an oil.

16

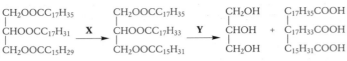

$$CH_2OOCC_{17}H_{35}$$
$$CHOOCC_{17}H_{31} \quad \mathbf{X} \quad$$
$$CH_2OOCC_{15}H_{29}$$

$$CH_2OOCC_{17}H_{35}$$
$$CHOOCC_{17}H_{33} \quad \mathbf{Y} \quad$$
$$CH_2OOCC_{15}H_{31}$$

$$CH_2OH \qquad C_{17}H_{35}COOH$$
$$CHOH \quad + \quad C_{17}H_{33}COOH$$
$$CH_2OH \qquad C_{15}H_{31}COOH$$

A Hydration	**B** Addition	**C** Hydrolysis
D Oxidation	**E** Hydrogenation	**F** Condensation

(a) Identify the name which could be applied to reaction **Y**.
(b) Identify the name(s) which could be applied to reaction **X**.

17 Carbohydrates are very important foodstuffs.

A	fructose
B	glucose
C	maltose
D	starch
E	sucrose

(a) Identify the **two** carbohydrates which do **not** give a positive result when tested using Benedict's or Fehling's Reagent.
(b) Identify the **two** carbohydrates which have the molecular formula $C_6H_{12}O_6$.

18

A O—H with C and benzene ring	**B** HOCH$_2$CH$_2$OH	**C** structure with C=O, HO, C, N, H, H, CH$_3$
D COOH / benzene ring / COOH	**E** C$_2$H$_5$NH$_2$	**F** O=C—CH$_3$ with benzene ring

(a) Identify the molecule which could be produced when a protein is hydrolysed.
(b) Identify the **two** molecules which contain the carboxyl group.
(c) Identify the **two** molecules which could be used to make polyester.

19 Fats and oils are naturally occurring esters.

Identify the **true** statement(s).

A	Fats and oils in the diet can supply the body with energy.
B	Fats and oils are a less concentrated source of energy than carbohydrates.
C	Fats are likely to have relatively low melting points compared to oils.
D	Fats are likely to have a higher degree of unsaturation than oils.
E	Fats and oils are esters.

20 There are many different types of reactions.

A Addition	**B** Condensation	**C** Dehydration
D Hydrogenation	**E** Hydrolysis	**F** Polymerisation

Poly(ethenol) is a recently developed plastic which is soluble in water. It is made by the reactions shown.

(a) Identify the type of reaction taking place at Step 2.
(b) Identify the term(s) which can be applied to the reaction taking place at Step 1.

21 Some water is added to hydrochloric acid, concentration $2 \text{ mol } l^{-1}$. Identify the **true** statement(s).

A	The pH increases
B	The acid is neutralised.
C	The concentration of $H^+(aq)$ ions increases.
D	The number of chloride ions in the solution decreases.
E	The volume of alkali needed for neutralisation decreases.
F	The speed of the reaction of the acid with zinc decreases.

22 Ann added 20 cm³ of 1 mol l⁻¹ sodium hydroxide solution to 20 cm³ of 1 mol l⁻¹ sulphuric acid.

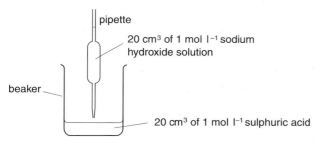

Identify the statement(s) which can be applied to this experiment.

A	The number of $H^+(aq)$ ions in the beaker decreased.
B	The pH of the solution decreased.
C	The number of $SO_4^{2-}(aq)$ ions in the beaker decreased.
D	Water molecules formed during the reaction.
E	A precipitate formed during the reaction.
F	The final solution contained equal numbers of $H^+(aq)$ and $OH^-(aq)$ ions.

23 Magnesium sulphate can be converted to magnesium nitrate in two stages as follows.

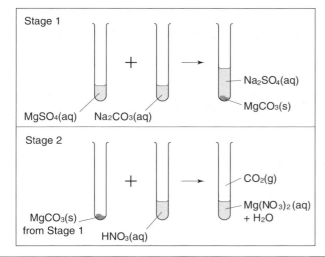

A	Reduction	B	Oxidation	C	Neutralisation
D	Precipitation	E	Addition	F	Displacement

(a) Identify the type of reaction involved in Stage 1.
(b) Identify the type of reaction involved in Stage 2.

24 Many pairs of chemicals react together.

A	sodium carbonate + ethanoic acid	B	silver + dilute hydrochloric acid
C	copper oxide + dilute sulphuric acid	D	sodium hydroxide solution + dilute nitric acid
E	sodium + water	F	zinc + copper sulphate solution

(a) Identify the pair(s) of chemicals which react to produce a gas.
(b) Identify the pair of chemicals which will **not** react.

25 Precipitation reactions can be used to prepare some salts.

A	copper nitrate	B	barium sulphate
C	lithium carbonate	D	sodium chloride

(a) Identify the salt which can be prepared by a precipitation reaction.
(b) Identify the **two** soluble salts whose solutions would form a precipitate when mixed.

26 Some 4th year pupils set up the following experiments to study the sacrificial protection of metals. Each tube contains ferroxyl indicator.

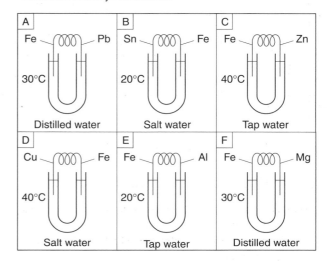

(a) Identify the **two** experiments which would give a fair comparison of the ability of different metals to protect iron.
(b) Identify the experiment in which the iron will rust most quickly.

27 Jane used different pairs of metals to produce electricity. She set up experiments as shown in the diagram.

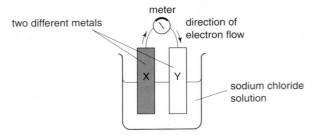

The table shows the different pairs of metals she used.

	X	Y
A	lead	zinc
B	tin	nickel
C	zinc	tin
D	lead	nickel

(a) Which pair(s) of metals would produce a flow of electrons in the direction shown in the diagram?

(b) Which pair of metals would produce the highest voltage?

28 Ion-electron equations can be used to show the gain and loss of electrons in chemical reactions.

A $Fe(s) \rightarrow Fe^{2+}(aq) + 2e$	**B** $Fe^{2+}(aq) + 2e \rightarrow Fe(s)$
C $Fe^{2+}(aq) \rightarrow Fe^{3+}(aq) + e$	**D** $Fe^{3+}(aq) + e \rightarrow Fe^{2+}(aq)$
E $Cu(s) \rightarrow Cu^{2+}(aq) + 2e$	**F** $Cu^{2+}(aq) + 2e \rightarrow Cu(s)$

(a) Identify the equation which shows iron(II) ions being oxidised.

(b) Identify the **two** equations which show the reactions which occur when an iron nail is placed in copper(II) sulphate solution.

29 Objects made of iron are often galvanised, i.e. coated with zinc.

Identify the statement(s) which can apply to a galvanised farm gate that has been scratched.

A	The zinc increases the rate of corrosion of iron.
B	The zinc is oxidised.
C	The zinc attracts electrons from the iron.
D	The zinc does not corrode.
E	The zinc corrodes slower than the iron.
F	The zinc is sacrificed to protect the iron.

30 Iron(III) chloride solution reacts with potassium iodide solution in a redox reaction.

$$2FeCl_3 + 2KI \rightarrow 2FeCl_2 + I_2 + 2KCl$$

The grid shows the ions present during the reaction.

A Fe^{3+}	B Cl^-	C K^+	D I^-	E Fe^{2+}

(a) Identify the ion which is reduced.

(b) Identify the **two** spectator ions in the reaction.

31 Identify the statement(s) which can be applied to iron but **not** to copper.

A	It displaces tin from a solution of tin chloride.
B	It reacts with cold water.
C	It can be obtained by heating its oxide with carbon.
D	It reacts with dilute hydrochloric acid.
E	It is displaced from a solution of its chloride by zinc.

32 Reactions can be represented using ionic equations.

A	$H^+(aq) + OH^-(aq) \rightarrow H_2O(l)$
B	$2H_2O(l) + O_2(g) + 4e \rightarrow 4OH^-(aq)$
C	$2H^+(aq) + CO_3^{2-}(aq) \rightarrow H_2O(l) + CO_2(g)$
D	$SO_2(g) + H_2O(l) \rightarrow 2H^+(aq) + SO_3^{2-}(aq)$
E	$NH_4^+(s) + OH^-(s) \rightarrow NH_3(g) + H_2O(l)$

(a) Identify the ionic equation representing reduction.

(b) Identify the ionic equation which shows the formation of acid rain.

(c) Identify the **two** ionic equations which show the neutralisation of an acid.

Part 2 Extended answer questions

1 Three experiments were carried out in a study of the rate of reaction between magnesium (in excess) and dilute hydrochloric acid. A balance was used to record the mass of the reaction flask and its contents. The results of Experiment 1, using 0.4 mol l^{-1} acid, are shown in the graph.

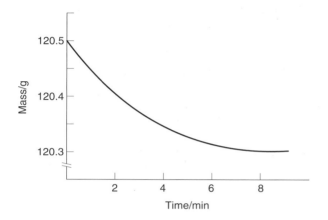

(a) Why did the balance record a decrease in mass during the reaction?
(b) The **only** difference between Experiment 2 and Experiment 1 was the use of a catalyst. Copy the graph and sketch a curve that could be expected for Experiment 2.
(c) The **only** difference between Experiment 3 and Experiment 1 was the use of 0.2 mol l^{-1} acid. Sketch a curve that could be expected for Experiment 3.

2 Find the missing terms in the following table:

Ion	Number of protons	Number of neutrons	Electron arrangement
$^{23}_{11}Na^+$	(a)	(b)	(c)
$^{40}_{20}Ca^{2+}$	(d)	(e)	(f)
(g)	13	14	2, 8

3 The graph shows annual sales of leaded and unleaded petrol from 1984 onwards.

(a) Describe the trends in the amounts of leaded and unleaded petrol sold between 1988 and 1994.
(b) Predict the sales of leaded and unleaded petrol for 1998 if these trends continue.
(c) Most modern car exhausts are fitted with a catalytic converter.

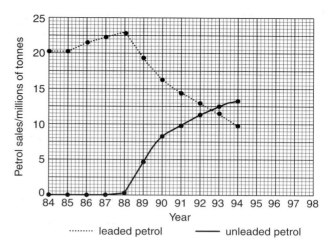

(i) What type of metal is used as the catalyst?
(ii) What does a catalytic converter do?
(d) Mixing more air with the petrol in an engine makes combustion more complete. This decreases pollution. Name a pollutant which is decreased.

4 Crude oil arriving at the BP/Amoco refinery in Grangemouth is separated into different fractions.

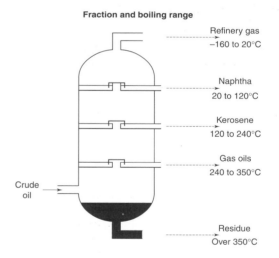

(a) In which fraction will pentane be found?
(b) Why is the gas oils fraction more viscous than the kerosene fraction?
(c) Fractions which are surplus to requirements can be cracked.
(i) Give a reason for cracking fractions.
(ii) A catalyst is used to speed up this process. Suggest another reason for using a catalyst.

5 The compound diazomethane, CH_2N_2, undergoes an unusual reaction called **insertion**. Under certain experimental conditions, the CH_2 group produced can insert itself into **any** bond which includes an atom of hydrogen.

eg

Nitrogen is a product in every reaction.

One of the products for the reaction of diazomethane with ethanol is shown below.

(a) Name the product shown.
(b) Draw the full structural formula for the other **two** organic products which could be formed in this reaction.

6 Laundry bags, used in hospitals, are made from poly(ethenol), a polymer which will dissolve in hot or cold water. Poly(ethenol) is made from a monomer which has the following structure.

(a) What type of polymerisation produces poly(ethenol)?
(b) Draw the full structural formula for a section of the polymer made from **three** of these monomer units.

7 Proteins are polymers found in the human body in tissues such as hair, finger nails and skin.

(a) Name the type of compound produced by hydrolysis of a protein.
(b) Part of a protein molecule is shown.

Draw the structural formula for **one** of the monomers produced by hydrolysis of this protein fragment.

8 Methanol and ethanol can be used as alternative fuels in car engines.

(a) Why is ethanol considered to be a 'renewable' fuel?
(b) To meet market demand, ethanol is also made by a method other than fermentation. What is this method?

9 Kevlar is a strong, low density, linear polymer. Part of the molecular structure is shown.

(a) What kind of polymerisation takes place in the formation of Kevlar?
(b) One of the monomers used to make Kevlar is shown.

Draw a structural formula for the other monomer.

10 Two reactions of propanoic acid are shown.

(a) Draw a structural formula for ester **X**.
(b) Give a name for compound **Y**, which reacts with propanoic acid to form sodium propanoate.

11 Proteins are polymers formed when amino acids react. Glycine is an example of an amino acid.

(a) Draw the part of the protein molecule which is formed by the three glycine molecules linking together.

(b) Name this kind of polymerisation.

12 Many of the foods we eat contain starch. The starch molecules are hydrolysed in our bodies.

(a) Why do starch molecules need to be hydrolysed in our bodies?
(b) Why do foods containing glucose give us energy faster than starch foods?

13 Fats and oils are naturally occurring esters of fatty acids and glycerol.

(a) (i) Why do oils tend to have lower melting points than fats?
(ii) Draw the full structural formula for glycerol.
(b) The full structural formula for an ester found in pineapples is shown.

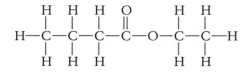

(i) Name the ester.
(ii) What would be observed if this ester was added to water?

14 Carbon monoxide can be used to extract iron from iron(III) oxide. Carbon dioxide is produced in the reaction.

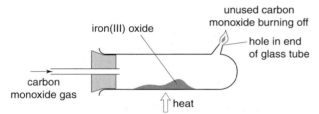

(a) Write a balanced equation for the reaction between iron(III) oxide and carbon monoxide.
(b) Name the type of reaction which takes place.
(c) Explain why calcium cannot be extracted from its oxide in this way.

15 Some hearing aid batteries contain silver(I) oxide.

Silver can be extracted from the silver(I) oxide in spent batteries, using a two-step process.

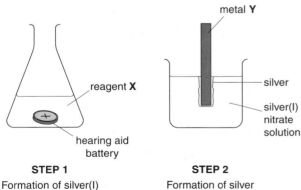

STEP 1
Formation of silver(I) nitrate solution

STEP 2
Formation of silver

(a) Name reagent **X**.
(b) The reaction in step 2 is an example of a redox reaction. Give another name for the kind of reaction occurring in this step.
(c) No silver is produced if metal **Y** is platinum. What does this indicate about the reactivity of platinum?

16 Iain determined the mass of iron present in a piece of wire.

STEP 1 He completely reacted the wire with dilute sulphuric acid. This produced a solution of iron(II) sulphate, $FeSO_4$

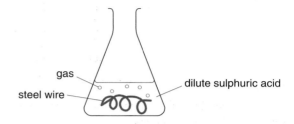

STEP 2 He found the volume of potassium permanganate solution needed to react with the iron(II) sulphate solution.

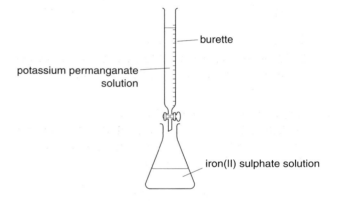

(a) (i) Why was the wire reacted with sulphuric acid in step 1?
(ii) Name the gas formed in step 1.
(b) Iain found that 20.2 cm^3 potassium permanganate solution, concentration of 0.15 mol l^{-1}, were needed to react with the iron(II) sulphate solution.
(i) How many moles of potassium permanganate were needed to react with the iron(II) sulphate?
(ii) One mole of potassium permanganate reacts with five moles of iron(II) sulphate. Calculate the mass of iron present in the steel wire.

17 Chromium(VI) oxide dissolves in water to form chromic acid.

$$CrO_3(s) + H_2O(l) \rightarrow H_2CrO_4(aq)$$

(a) Suggest what is unusual about the reaction between chromium(VI) oxide and water.
(b) Chromic acid forms a salt with potassium hydroxide.
(i) How many moles of potassium hydroxide would be needed to neutralise completely one mole of chromic acid?
(ii) Name the salt formed in the reaction.

18 This graph shows how much lime a gardener needs to add to produce a soil of the desired pH, e.g. if the soil had a pH of 4 but a pH of 7 was required, 1.2 kg of lime would be needed for every 10 m² of soil.

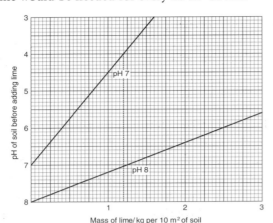

y-axis: pH of soil before adding lime
x-axis: Mass of lime/ kg per 10 m² of soil
(labels on graph: pH 7, pH 8)

(a) What mass of lime is needed to increase the pH of soil in a 100 m² garden from pH of 6 to pH of 8?
(b) The formula for the lime used is $Ca(OH)_2$. Calculate the number of moles of lime in 1.2 kg.

19 (a) Calculate the mass of sodium hydroxide which would be required to make 1 litre of exactly 1 mol l^{-1} solution.
(b) Solid sodium hydroxide often contains sodium carbonate as an impurity. This is caused by some of the sodium hydroxide reacting with carbon dioxide from the air.

INSTRUCTION CARD

Preparation of approximately 1 mol l⁻¹ sodium hydroxide solution

CARE: Sodium hydroxide is highly corrosive

1. Weigh out 41g of sodium hydroxide pellets.
2. Dissolve in about 400 cm³ of water.
3. Add a few drops of barium chloride solution to remove the carbonate ions.
4. Filter.
5. Make the volume of solution up to exactly 1 litre.

The equation for Step 3 is:

$$BaCl_2(aq) + Na_2CO_3(aq) \rightarrow BaCO_3(s) + 2NaCl(aq)$$

(i) Name an ion which will be an impurity in the final sodium hydroxide solution.
(ii) Rewrite the equation as an ionic equation omitting spectator ions.
(iii) 25.0 cm³ of the solution was neutralised by 25.5 cm³ of 1 mol l^{-1} hydrochloric acid. Calculate the exact concentration of the sodium hydroxide solution.

20 Common salt is extracted from salt mines by pumping in water to dissolve it.

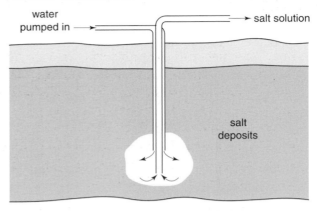

water pumped in → / → salt solution / salt deposits

The salt solution which is obtained contains magnesium sulphate as an impurity.

The magnesium ions are removed by adding sodium hydroxide solution.

$$Mg^{2+}(aq) + SO_4{}^{2-}(aq) + 2Na^+(aq) + 2OH^-(aq) \rightarrow$$
$$Mg^{2+}(OH^-)_2(s) + 2Na^+(aq) + SO_4{}^{2-}(aq)$$

(a) (i) Name the type of reaction taking place between the magnesium sulphate solution and the sodium hydroxide solution.
    (ii) Name the spectator ions in the reaction.
(b) The products of the reaction can be separated by filtration. Draw and label the apparatus you would use in the laboratory to carry out this process.

Indicate on the diagram where each product is collected.

21 Silver oxide cells are used in calculators and watches.

(a) (i) How is electricity produced in a cell?
    (ii) Give a disadvantage of a cell.

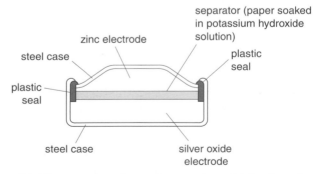

(labels: separator (paper soaked in potassium hydroxide solution), zinc electrode, steel case, plastic seal, plastic seal, steel case, silver oxide electrode)

(b) The equation shows the reaction which takes place at the zinc electrode.

$$Zn(s) + 2(OH)^-(aq) \rightarrow ZnO(s) + H_2O(l) + 2e$$

State why the equation represents oxidation.
(c) Why has the separator been soaked in potassium hydroxide solution?

20 Questions Involving Prescribed Practical Activities

1 In an experiment to study the effect that varying the concentration of a reactant has on reaction rate, a solution containing sodium persulphate and starch was mixed with a solution containing potassium iodide and sodium thiosulphate. Persulphate ions and iodide ions react to produce iodine which, in turn, is converted back to iodide ions by reaction with thiosulphate ions.

(a) What sharp change in colour of the solution marks the point when all of the thiosulphate ions have been used up.

(b) In each experiment the time (t) from the mixing of the solutions until the sharp change in colour is noted. The results of one set of experiments were used to plot a graph of reaction rate (l/t) against the relative concentration of sodium persulphate solution which is shown below.

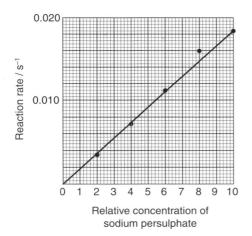

Relative concentration of
sodium persulphate

An error occurred when calculating the reaction rate for a relative concentration of 8 for the sodium persulphate.

(a) Based on the graph, what should the reaction rate have been for this experiment?

(b) Calculate the time taken from the mixing of the solutions until the sharp colour change.

(c) Based on the graph, what conclusion can be reached concerning the reaction rate and the concentration of the sodium persulphate solution?

(d) When carrying out the experiments the total volume of the solutions reacting was kept constant. Name *two* other possible variables which must be kept constant throughout the experiments.

2 The reaction between sodium thiosulphate solution and dilute hydrochloric acid can be used to investigate the effect of change in temperature on reaction rate. During the investigation the reaction mixture becomes more cloudy due to the formation of sulphur; the time taken for a fixed mass of sulphur to form is noted.

(a) State how the experiment is carried out in order to determine the time taken for a fixed mass of sulphur to form.

(b) During a series of experiments, each carried out using a different temperature for the reaction mixture, the following results were obtained:

Temperature/°C	Time (t) for formation of a fixed mass of sulphur/s	Reaction rate (l/t)/s^{-1}
10	125	0.008
20	77	0.013
30	44	0.023
40	26	0.038
50	X	0.059

Draw a line graph of reaction rate against temperature.

(c) Calculate the value for X in the table.

(d) State four factors, concerning *only* the reacting solutions, which must be kept constant throughout the investigation.

3 When copper(II) chloride solution is electrolysed, chemical reactions take place at the electrodes.

(a) Draw a labelled diagram of the apparatus used.

(b) State what is observed at the negative electrode.

229

(c) i) What is formed at the positive electrode?
   ii) Describe the chemical test that you would carry out at the positive electrode in order to confirm your answer to (c)i). Give the results of your test.
(d) Describe a hazard which is associated with the product formed at the positive electrode and a control measure which should be adopted in order to reduce it.

4 A hydrocarbon burns with a smoky yellow flame suggesting a high carbon content and that the hydrocarbon may be unsaturated.

(a) What is meant by the term *unsaturated* in this context?
(b) Describe how the test for unsaturation is carried out and give the result of the test if the substance being tested is in fact unsaturated.
(c) A hydrocarbon X, with molecular formula C_4H_8, proved to be unsaturated. Draw two possible full structural formulae for this compound and give names for each.
(d) A hydrocarbon Y, with molecular formula C_4H_8, proved to be saturated. Draw a possible full structural formula for this compound and give its name.
(e) Describe any hazard that is associated with the reagent used in the test for unsaturation and a control measure which can be used in order to reduce it.

5 The apparatus used in an experiment involving the cracking of the mixture of long chain alkanes called 'liquid paraffin' is shown below. The object of the experiment is to show that the mixture of compounds produced contains unsaturated hydrocarbons.

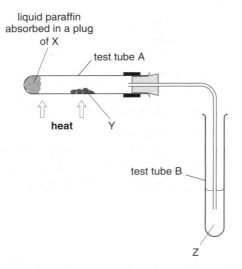

(a) State what X, Y and Z are in the diagram.
(b) State what is meant by the term 'cracking'.
(c) A catalyst can be used to speed up the reaction, but that is not the reason why a catalyst is used in cracking reactions. State the correct reason.
(d) What colour change is observed in the test tube containing Z when the products of cracking enter it?
(e) Why must the delivery tube be removed from the liquid in test tube B *before* the heating of test tube A is stopped?
(f) One of the alkanes present in liquid paraffin is octadecane, $C_{18}H_{38}$. If on cracking this breaks into two hydrocarbons, one of which is oct–1–ene, which has molecular formula C_8H_{16}, give the name and molecular formula for the other compound formed.
(g) Why is it not possible, during the cracking of a large alkane molecule, to obtain two smaller alkane molecules as the only products?

6 The diagrams below show two ways of hydrolysing starch.

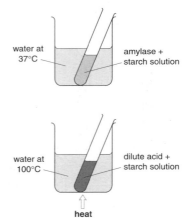

(a) Describe the chemical test that would show the presence of starch. Give the result of the test.
(b) In both cases, starch undergoes a hydrolysis reaction to form glucose. What is meant by the term 'hydrolysis reaction'?
(c) Describe a chemical test which would confirm that glucose, or a similar sugar, had been produced. Describe the colour change during the test.
(d) Name the enzyme that is used to hydrolyse starch.
(e) Explain why it is important to also have a control test tube containing only starch solution in each beaker.

7 Crystals of magnesium sulphate can be prepared from the insoluble base magnesium carbonate and dilute sulphuric acid.

(a) Name all the products of the reaction between magnesium carbonate and dilute sulphuric acid.

(b) State how crystals of magnesium sulphate can be prepared using magnesium carbonate and 20 cm³ of dilute sulphuric acid, dealing in turn with the following procedures:
i) neutralisation (indicating *three* ways of knowing that neutralisation is complete)
ii) filtration
iii) evaporation
iv) crystallisation

(c) Explain what safety precaution would be needed if magnesium sulphate was made from magnesium and dilute sulphuric acid rather than magnesium carbonate and the same acid.

8 Various factors *might* affect the voltage of a simple cell, such as the one shown below.

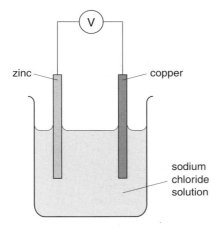

One factor which affects the voltage is the use of different pairs of metals. In one set of experiments, the following readings were recorded:

Metal A	Metal B	Voltage/V
iron	copper	0.5
aluminium	copper	1.3
zinc	copper	0.9

(a) How is the voltage produced related to the relative positions of the metals concerned in the electrochemical series?

(b) Although correctly positioned by the voltage obtained, a higher result was obtained when aluminium was rubbed with sandpaper before being tested. What thin protective coating would be removed by the sandpaper, thus allowing better contact with the metal?

(c) State three factors which should be kept constant when carrying out the experiments to see whether the use of different pairs of metals affects the voltage produced in a cell.

9 Some metals can be placed in order of reactivity by observing their reaction when heated in oxygen. The apparatus which is normally used is shown below.

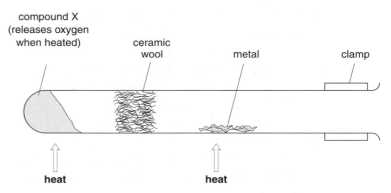

(a) Name the compound X which releases oxygen when heated.

(b) Why is it important to heat the metal *first* before heating compound X to release oxygen?

(c) In one set of experiments the following results were obtained:

Metal	Observation
copper	changed from pink/brown to black, no glow
magnesium	
iron	no significant colour change, glowed brightly

State the observations that you would expect to be made for magnesium, making reference to the likely vigour of reaction and to the change in appearance.

(d) In order that the comparisons are fair, state two factors which must be kept constant relating to the metals themselves.

 21 Chemical Dictionary

A

acid   A compound which produces $H^+(aq)$ ions when it dissolves in water.

acid rain   Unusually acidic rain caused by the presence of sulphuric and nitric acids.

acidic solution   One which contains a greater concentration of $H^+(aq)$ ions than pure water.

addition polymerisation   A reaction in which many small monomer molecules join to form one large polymer molecule and nothing else.

addition reaction   One in which two or more molecules join to produce a single larger molecule and nothing else.

adsorption   The attachment of molecules of a liquid or gas to the surface of another substance (usually a solid).

alkali   A compound which produces $OH^-(aq)$ ions when it dissolves in water.

alkaline solution   One which contains a greater concentration of $OH^-(aq)$ ions than pure water.

alkali metals   Reactive metals in group 1 of the Periodic Table (Li, Na, K, Rb, Cs, Fr).

alkanes   A homologous series with general formula C_nH_{2n+2}. The simplest is methane, CH_4.

alkanoic acids   A homologous series in which a $-CH_3$ group in an alkane has been replaced by a carboxyl group $-COOH$. The simplest is methanoic acid HCOOH.

alkanols   A homologous series in which a hydrogen atom in an alkane has been replaced by a hydroxyl group, $-OH$. The simplest is methanol, CH_3OH.

alkenes   A homologous series with general formula C_nH_{2n}. The simplest is ethene, C_2H_4.

amide link   The group $-CONH-$ present in polyamides. (Same group of atoms as in the peptide link.)

amine group   The group $-NH_2$ present in, for example, amines and amino acids. Also called the amino group.

amino acid   Compounds containing the amine group $-NH_2$ and the carboxyl group $-COOH$. They undergo condensation polymerisation in the formation of proteins.

amino group   See **amine group**.

amines   Compounds in which a hydrogen atom in an alkane, or other hydrocarbon, has been replaced by an amine group $-NH_2$.

amylase   The enzyme which catalyses the hydrolysis of starch to glucose in the body.

atom   The smallest part of an element that can exist. It has a nucleus of protons and (usually) neutrons surrounded by moving electrons.

atomic number   The number of protons in the nucleus of an atom.

B

balanced chemical equation   One with the same number of atoms of each element on both sides of the equation.

base   A substance which can accept hydrogen ions (from an acid), e.g. ammonia and oxides, hydroxides and carbonates of metals.

battery   A series of chemical cells joined together. (Often the words cell and battery are used interchangeably.)

Benedict's solution   A solution which turns from blue to orange-red when heated with many sugars, e.g. glucose, fructose and maltose (but not sucrose).

biodegradeable   Able to rot away by natural biological processes.

burning   A chemical reaction in which a substance combines with oxygen releasing energy in the form of heat and light.

C

carbohydrates   Group of compounds with general formula $C_x(H_2O)_y$. Important class of foods made by plants.

carboxylic acid   One containing the carboxyl group $-COOH$.

carboxyl group   The group $-COOH$ found in carboxylic acids such as the alkanoic acids.

catalyst   A substance which speeds up a chemical reaction, but is not used up and can be recovered unchanged at the end of the reaction.

catalyst poison   Something which destroys the activity of a catalyst.

catalytic converter   Part of the exhaust system of a car where platinum and other catalysts change the pollutants carbon monoxide, oxides of nitrogen and unburned hydrocarbons into less harmful carbon dioxide, nitrogen and water vapour.

catalytic hydration   An addition reaction in which water adds on to a molecule in the presence of a catalyst, e.g. ethene and water produce ethanol.

cell   In a *chemical* cell chemical energy is converted into electrical energy. In an *electrolytic* cell, electrical energy is used to bring about chemical reactions at the electrodes.

chemical equation   See **formulae equation**.

chemical reaction   A chemical process in which one or more new substances are formed (usually accompanied by an energy change and a change in appearance).

chromatography   A method for separating small quantities of soluble compounds. (Refers to *paper* chromatography.)

combustion   See **burning**.

compound   A substance in which two or more elements are joined together chemically.

concentrated solution   One with a lot of solute compared to the amount of solvent.

concentration   The amount of solute in a given volume of solution. The usual units are moles per litre (mol l^{-1}).

condensation polymerisation   A reaction in which many small monomer molecules join to form one large polymer molecule with water, or some other small molecule, formed at the same time.

condensation reaction   One in which two or more molecules join to produce a single larger molecule, with water or another small molecule formed at the same time.

corrosion   A chemical reaction in which the surface of a metal changes from an element to a compound.

covalent bond   The result of two nuclei being held together by their common attraction for a shared pair of electrons.

covalent network   A giant lattice of atoms held together by covalent bonds, e.g. diamond, graphite and silicon dioxide.

cracking   The breaking up of larger hydrocarbon molecules (usually alkanes) to produce a mixture of smaller molecules (usually alkanes and alkenes). The use of a catalyst (in catalytic cracking) allows the process to take place at a lower temperature.

crude oil   A naturally occurring mixture consisting mainly of hydrocarbons. An alternative name is petroleum.

cycloalkanes   A homologous series of ring molecules with general formula C_nH_{2n}. The simplest is cyclopropane, C_3H_6.

D

decomposition   The breaking down of a compound into two or more substances (usually by heating), e.g. $2HgO \rightarrow 2Hg + O_2$

diamines   Compounds in which two hydrogen atoms in an alkane, or other hydrocarbon, have both been replaced by amine groups, $-NH_2$.

diatomic element   Element whose molecules contain only two atoms, e.g. H_2, N_2, O_2, F_2, Cl_2, Br_2, I_2 and At_2.

diatomic molecule   Molecule which contains only two atoms, e.g. H_2, HCl and CO.

dilute solution   One with little solute compared to the amount of solvent.

displacement reaction   Formation of a metal from a solution containing its ions by reaction with a metal higher in the ECS.

distillation   A process of separation or purification dependent on differences in boiling point. The changes of state involved are:

$$liquid \rightarrow gas \rightarrow liquid$$

E

electrochemical series (ECS)   A list of metals (and hydrogen) in order of their ability to lose electrons and form ions in solution. (Similar to the reactivity series of metals.)

electrode   A conductor which is used to pass electricity into and out of solutions and melts, e.g. carbon as graphite.

electrolyte   A compound which conducts due to the movement of ions either when dissolved in water or melted.

electrolysis   The flow of ions through electrolytes (ionic solutions or molten ionic compounds) accompanied by chemical changes at the electrodes which can result in the decomposition of the electrolyte.

electron   A particle which moves around the nucleus of an atom. It has a single negative charge but its mass is negligible compared to that of a proton or neutron.

electroplating   A process in which a layer of metal is deposited on another metal by electrolysis. The latter metal is used as the negative electrode in a solution containing ions of the metal being deposited.

element   A substance which cannot be broken down into simpler substances by chemical means. All of its atoms have the same atomic number.

empirical formula   Shows the simplest ratio of atoms in a compound, e.g. SiO_2 is the empirical formula for silicon dioxide.

end point   The point of a volumetric titration when the reaction is complete, which is shown by a colour change in a suitable indicator.

endothermic reaction   One in which heat energy is taken in from the surroundings causing a decrease in temperature.

enzyme   A biological catalyst.

equilibrium   A situation when forward and reverse reactions in a reversible reaction take place at the same rate. The concentrations of reactants and products are constant but not necessarily equal.

ester group   The group $-COO-$ found in esters.

esters   Compounds formed on reaction between an alkanol and an alkanoic acid. The simplest is methyl methanoate, $HCOOCH_3$.

evaporation   The process of turning a liquid into a vapour (gas), usually by heating.

exothermic reaction   One in which heat energy is given out to the surroundings causing an increase in temperature.

F

fats and oils   An important group of high energy foods. They are esters of glycerol and fatty acids.

fatty acids   Saturated or unsaturated straight chain carboxylic acids, usually with long chains of carbon atoms, e.g. stearic acid, $C_{17}H_{35}COOH$.

fermentation   The breakdown of glucose to produce ethanol and carbon dioxide catalysed by enzymes in yeast. (Can be applied to the similar breakdown of other organic molecules.)

ferroxyl indicator   Used to show the production of Fe^{2+}(aq) ions (by the formation of a blue colour) and OH^-(aq) ions (by the formation of a pink colour) during rusting of iron experiments.

filtrate The liquid which passes through filter paper during filtration.

filtration The separation of an insoluble solid from a liquid by passage through filter paper.

finite Limited.

flame A region where gases combine producing heat and light energy.

flammable Describes a substance that is easily set on fire.

formula A simple chemical formula for a molecular substance usually indicates the number of atoms of each element in a molecule, e.g. CO_2. In other cases the ratio of atoms or ions present is shown, e.g. NaCl. (See also **empirical, ionic, molecular, perspective** and **structural formulae**.)

formulae equation Equation which gives chemical symbols and formulae for reactants and products.

formula mass (more correctly **relative formula mass**) The sum of the relative atomic masses of all the atoms present in a formula.

fossil fuel One which has been formed from the remains of living things, e.g. coal, crude oil and natural gas.

fraction A group of hydrocarbons with boiling points within a given range obtained from crude oil by fractional distillation.

fractional distillation A means of separating crude oil into groups of hydrocarbons with similar boiling points (fractions).

fuel A substance which burns to produce energy.

functional group A group of atoms with characteristic chemical activity, e.g. carboxyl group −COOH.

G

galvanising Process by which iron is coated is coated with a protective layer of zinc (by dipping into molten zinc).

general formula One from which the molecular formula for any member of a homologous series can be deduced, e.g. C_nH_{2n+2} for the alkanes.

glycerol An alcohol with three −OH groups in each molecule, $CH_2OHCHOHCH_2OH$. Produced by the hydrolysis of fats or oils.

greenhouse gas One that causes global warming, e.g. carbon dioxide, CO_2.

group A column of elements in the Periodic Table.

group ion One which contains more than one kind of atom, e.g. the ammonium ion, NH_4^+.

H

halogens Reactive non-metals in group 7 of the Periodic Table (F, Cl, Br, I, At).

heterogeneous catalyst One which is in a different state from the reactants.

homogeneous catalyst One which is in the same state as the reactants.

homologous series A group of chemically similar compounds which can be represented by a general formula. Physical properties change gradually through the series, e.g. the alkanes.

hydrocarbon A compound that contains carbon and hydrogen only.

hydrolysis reaction One in which a large molecule is broken down into two or more smaller molecules by reaction with water.

hydroxyl group The group −OH found, for example, in water and alkanols.

I

indicator A substance whose colour is dependent on pH.

iodine solution A solution which turns from reddish-brown to dark blue in the presence of starch.

ion   Atom or group of atoms which possess a positive or negative charge due to loss or gain of electrons, e.g. Na^+ and CO_3^{2-}.

ionic bond   The electrostatic force of attraction between oppositely charged ions.

ionic equation   One which shows any ions that may be present among the reactants and products.

ionic lattice   A giant arrangement of ions held together by electrostatic attraction (ionic bonds), e.g. sodium chloride, NaCl.

isomers   Compounds which have the same molecular formula but different structural formulae.

ion bridge   Used to complete the circuit in a chemical cell by allowing a flow of ions through it.

ion-electron equation   One which shows either the loss of electrons (oxidation) or the gain of electrons (reduction).

isotopes   Atoms of the same element which have different numbers of neutrons. They have the same atomic number but different mass numbers.

L

lime water   Calcium hydroxide solution. It is used to test for carbon dioxide, which turns it milky white.

M

malleable   Describes a material which can be shaped by hammering or rolling – a physical property of metals.

mass number   The total number of protons and neutrons in the nucleus of an atom.

metals   Shiny, malleable elements found on the left of the Periodic Table. They all conduct electricity.

metallic bond   The electrostatic force of attraction between positive ions in a metallic lattice and the delocalised electrons within the structure.

mixture   Two or more substances mixed together but not chemically joined, e.g. air and sea water.

mole   A formula mass expressed in grams.

molecular formula   Formula which shows the number of atoms of different elements present in one molecule of a substance.

molecule   A group of atoms held together by covalent bonds.

monomer   Relatively small molecules which can join together to produce a very large molecule (a polymer) by a process called polymerisation.

N

natural gas   Gas obtained from underground deposits, which consists mainly of methane. When found in association with crude oil it contains appreciable quantities of other hydrocarbons such as ethane, propane and butane.

neutral solution   One in which the concentrations of $H^+(aq)$ and $OH^-(aq)$ ions are equal. The pH of a neutral solution is 7.

neutralisation reaction   A reaction of acids with bases in which the pH of a solution moves towards 7.

neutron   A particle found in the nucleus of an atom. It has the same mass as a proton but no charge.

noble gases   Very unreactive gases in group 0 of the Periodic Table (He, Ne, Ar, Kr, Xe, Rn).

non-metals   Non-conducting elements (except carbon as graphite) found on the right of the Periodic Table.

nucleus   The extremely small centre of a atom where the protons and neutrons are found.

O

oxidation reaction   One in which electrons are lost.

P

peptide link   The group $-CONH-$ present in proteins. (Same group of atoms as in the amide link.)

period A horizontal row in the Periodic Table.

Periodic Table An arrangement of the elements in order of increasing atomic number, with chemically similar elements occurring in the same vertical columns (groups).

perspective formula Formula which shows the true shape of a molecule.

petroleum Another name for crude oil.

pH A number which indicates the degree of acidity or alkalinity of a solution. Acidic solutions $pH < 7$; neutral solutions $pH = 7$; alkaline solutions $pH > 7$.

polar covalent bond Covalent bond in which the electron pair is not shared equally, making one atom slightly positive ($\delta+$) and the other slightly negative ($\delta-$).

pollutant Something which harms the environment.

polyamides Condensation polymers which can be made from dicarboxylic acids and diamines. They include the compounds collectively referred to as nylon.

polymer A very large molecule formed by the joining of many small molecules called monomers.

polymerisation The process whereby many small monomer molecules join to form one large polymer molecule. (See **addition polymerisation** and **condensation polymerisation**.)

precipitate An insoluble solid which is formed on mixing certain solutions.

precipitation reaction One in which two solutions react to form an insoluble solid called a precipitate.

proteins An important class of foods made by plants. All contain C, H, O and N and are condensation polymers made of many amino acids joined together.

proton A particle found in the nucleus of all atoms. It has a single positive charge and the same mass as a neutron.

R

reactivity series A list of metals in order of reactivity, e.g. with oxygen, water and dilute acids.

redox reaction One in which reduction and oxidation take place. Electrons are lost by one reactant and gained by another.

reduction reaction One in which electrons are gained.

relative atomic mass The average mass of one atom of an element on a scale where one atom of ^{12}C has a mass of 12 units exactly. It is the average of the mass numbers of an element's isotopes taking into account the proportions of each.

renewable energy source One which can be renewed and will not therefore run out in the foreseeable future.

residue The solid left behind in the filter paper after filtration.

reversible reaction One which proceeds in both directions, e.g.
alkanol + alkanoic acid $\rightleftharpoons$ ester + water.

rust The name of the compound formed when iron corrodes. It can be looked upon as iron(III) hydroxide or as a form of iron(III) oxide.

rusting The corrosion of iron. Both water and oxygen are needed for iron to rust.

S

sacrificial protection A method for protecting a metal from corrosion by attaching to it a metal which is higher in the ECS.

salt A compound in which the hydrogen ions of an acid have been replaced by metal ions or ammonium ions.

saturated hydrocarbon One in which all carbon-to-carbon covalent bonds are single bonds, i.e. C–C.

saturated solution A solution in which no more solute will dissolve at a given temperature.

solute A substance that dissolves in a liquid.

solution   A liquid mixture formed when a solute dissolves in a solvent.

solvent   A liquid in which a substance dissolves.

spectator ion   An ion which is present in a reaction mixture but takes no part in the reaction.

standard solution   One whose concentration is known accurately.

starch   A large carbohydrate molecule which is a natural condensation polymer made from glucose monomers.

state symbols   Symbols used to indicate the state of a substance or species: (s) = solid; (l) = liquid; (g) = gas; (aq) = aqueous (dissolved in water).

strong acid   One which dissociates (breaks up) completely in aqueous solution to produce ions, e.g. hydrochloric, nitric and sulphuric acids.

strong alkali   One which dissociates completely in aqueous solution to produce ions, e.g. sodium hydroxide and all other soluble metal hydroxides.

structural formula   One which shows the arrangement of atoms in a molecule or ion. A *full structural formula* shows all of the bonds,

e.g.

$$\begin{array}{ccccccc} & H & & H & & H & \\ & | & & | & & | & \\ H- & C & - & C & - & C & -H \quad \text{(propane)} \\ & | & & | & & | & \\ & H & & H & & H & \end{array}$$

A *shortened structural formula* shows the sequence of groups of atoms, e.g. $CH_3CH_2CH_3$ (propane).

sugars   Small carbohydrate molecules, e.g. fructose and glucose, both $C_6H_{12}O_6$, sucrose and maltose, both $C_{12}H_{22}O_{11}$.

T

thermoplastic   A plastic which softens on heating and can be reshaped, e.g. PVC and polythene.

thermosetting plastic   One which does not soften on heating, e.g. Bakelite and formica.

titration   See **volumetric titration**.

toxic   Poisonous.

U

universal indicator solution   A solution containing several indicators which can be used to give the approximate pH of a solution. It gives a range of colours depending on the pH of the solution. (Also called pH indicator solution.)

unsaturated hydrocarbon   One in which there is at least one carbon-to-carbon double covalent bond, C=C.

V

valency   A number which indicates the 'combining power' of atoms or ions.

vaporisation   The process in which a liquid (or solid) changes into a vapour (gas).

variable   Something that can be changed in a chemical reaction, e.g. temperature, particle size and concentration.

viscosity   A description of how 'thick' a liquid is, e.g. engine oil is 'thicker' (more viscous) than petrol.

volatile   Describes a liquid with a low boiling point which easily evaporates.

volumetric titration   An experiment in which volumes of reacting liquids are measured. In acid/alkali titrations it is usual practice to use a pipette for measuring out the alkali and to add the acid from a burette.

W

weak acid   One which dissociates (breaks up) only partially in aqueous solution to produce ions, e.g. ethanoic acid and all other carboxylic acids.

weak alkali   One which dissociates only partially in aqueous solution to produce ions, e.g. ammonia solution.

word equation   An equation which gives only the names of reactants and products.